Marine Bioinvasions: Patterns, Processes and Perspectives

Marine Bioinvasions: Patterns, Processes and Perspectives

Edited by

Judith Pederson
Massachusetts Institute of Technology, Sea Grant College Program, Cambridge, Massachusetts, USA

Reprinted from *Biological Invasions*, Volume 5 (1–2), 2003

Kluwer Academic Publishers
DORDRECHT / BOSTON / LONDON

A C.I.P. catalogue record for this book is available from the Library of Congress

ISBN 1-4020-1449-X

Published by Kluwer Academic Publishers,
P.O. Box 17, 3300 AA Dordrecht, The Netherlands

Sold and distributed in North, Central and South America
by Kluwer Academic Publishers,
101 Philip Drive, Norwell, MA 02061, USA

In all other countries, sold and distributed
by Kluwer Academic Publishers,
P.O. Box 322, 3300 AH Dordrecht, The Netherlands

Cover photo: Phyllorhiza punctata: a medusa approximately 50 cm
in bell diameter swimming at the sea surface. From Graham et al., p. 56.

Printed on acid-free paper

Printed in the Netherlands

Table of contents

Biological Invasions **5:** 1–2, 2003.

Introduction

This issue of *Biological Invasions* represents a collection of papers presented at the Second International Conference on Marine Bioinvasions, held in New Orleans, Louisiana, USA, April 9–11, 2001. The conference offered an opportunity to explore the ongoing marine invasion research and its application to policy and management issues. Many of the questions explored at the conference are universal to invasion science in all ecosystems. Are some habitats more easily invaded than others? Can we predict which organisms will become invasive? What are the impacts of invasive species on ecosystems? Can we use our knowledge to prevent, eradicate, control or manage invasions?

The papers in this issue reflect our growing understanding of the effects of introduced species and gaps in our knowledge. Issues that were the focus of the meeting included ecological impacts, ecological and economic implications, evolutionary consequences, patterns of invasions, and the feasibility of biological control in marine systems. Some of the underlying themes were the need for communication of scientific knowledge to policy makers and development of predictive models to assist with decision-making.

Several papers examine ecological impacts by introduced species through experimental studies and observation of changes in the field. Ross, Johnson and Hewitt use a variety of methods to examine the impacts of the predatory seastar (*Asterias amurensis*) on the benthos. Because pre-invasion studies were not available, the use of several approaches to evaluate ecosystem effects has provided insights that a single method alone could not provide. Whitlow, Rice and Sweeney examine the native soft-shelled clam's (*Mya arenaria*) behavioral response to burrow deeper when exposed to a predatory, non-native crab (*Carcinus maenas*) and identify implications for restoration efforts. Hunt and Behrens Yamada use field and laboratory studies to identify the effects of a native crab's (*Cancer productus*) predation on the European green crab (*Carcinus maenas*), and the latter's restriction to less desirable habitats.

Microbial introductions and their potential impacts occur at a smaller scale but may have broad ecological impacts. Hahn finds that the rate of decomposition of native eelgrass (*Zostera marina*) was lower than that of the non-native eelgrass (*Zostera japonica*), even though microbial abundance was not significantly different. As *Z. japonica* spreads and increases in abundance, these results imply increased nutrient cycling and changes in primary and secondary productivity. Graham and colleagues review the invasion of a tropical Pacific jellyfish (*Phyllorhiza punctata*) in the Gulf of Mexico that caused problems for shrimp fisheries by clogging nets and preying on commercially important eggs and larvae. The early assessment of this introduction provides a basis for future studies and examination of effects on the ecosystem.

Nearly all studies analyzing ecological impact research rely on information collected after an invasion has occurred. Until we have improved pre-invasion data, the recommendation to use multiple approaches to assess invasion consequences expands the robustness of the information and its application to policy decisions.

Evolutionary change is usually reflected through the apparently slow morphological modifications of lineages in the fossil record. Molecular studies now assist in detecting subtler and often faster changes. Rapid evolution may occur with some species – microbes and those that are reproductively compatible such as species of the fish genus *Tilapia*. Costa-Pierce provides an overview of tilapia in culture and in the wild that examines genetic similarities on a worldwide basis. His study shows that introduced exotic tilapia are reproductively compatible with feral tilapia. The gene pool of feral, naturalized and aquaculture fish represents high heterozygosity and is indicative of accelerated evolutionary change. His findings are particularly relevant in states that ban importation of some tilapia but may not recognize the alleles present in hybrid fish.

As we try to discern patterns and processes of invasions, we rely on historical documentation and fossil records to determine human-mediated introductions, natural range expansions, and habitats that are more vulnerable than others. Lee II, Thompson and Lowe re-examined the extensive data of benthic communities in San Francisco Bay to identify patterns of non-indigenous taxa and associations with different habitat types and at different scales. The history of invasions in the Caspian Sea using fossil records and molecular markers are the basis of the work by Grigorovich, Therriault and MacIsaac. Faunal changes from the late Miocene through the Pliocene to the Pleistocene show low genetic divergence in the Caspian and Black Seas Basins, suggesting natural and probably multiple colonizations during this time. This is contrasted with a 1,800-fold increase in rate of establishment of new species during the 20th century. The establishment of these species is attributed to human related activities, particularly shipping but also to deliberate introductions. The authors' examination of past records illustrates the challenges facing those attempting to identify changes over time, namely the lack of early identification of introductions and their sources. The benefits of new molecular approaches assist interpretation of data.

Biological control is a controversial method for managing pests and usually involves introduction of a nonindigenous species to control another nonindigenous species. This was one of the themes of the conference and two papers elaborate on this issue. Secord reviews biological control in terrestrial agroecosystems and extracts lessons learned for marine ecosystems. Although biological control is successful in controlling the targeted pests, it often has unwanted consequences for native species. Several marine, coastal, estuarine and brackish water ecosystems have biological control programs at different stages of planning including castration, predation, herbivory, and microbial control. For marine ecosystems, alternative approaches to biological control should be employed, such as prevention, early detection, and eradication of small, new populations, and use of native enemies. Kuris describes a biological control example that caused extinction of a native coconut moth, *Levuana iridescens* and possible collateral damage of other native species. The coconut moth was destroying the coconut palm, a source of income for the Fijian Archipelago. Its eradication has social, economic, and ecological impacts and highlights the multifaceted issues presented in attempting to balance management of ecosystems and human values.

As we look towards the future, we will continue to seek answers to the questions of what will become invasive, why one region and not another appears to be more easily invaded, and how can we use new knowledge to better prevent, control, and manage invasions in the world's oceans.

Judith Pederson
MIT Sea Grant College Program
Massachusetts Institute of Technology
Cambridge, MA 02139
USA
E-mail: jpederso@mit.edu

Biological Invasions **5:** 3–21, 2003.

Assessing the ecological impacts of an introduced seastar: the importance of multiple methods

D. Jeff Ross[1–3,*], Craig R. Johnson[1] & Chad L. Hewitt[2,4]
[1]*School of Zoology, University of Tasmania, Sandy Bay, Tasmania 7000, Australia;* [2]*Centre for Research on Introduced Marine Pests, CSIRO Marine Research, Hobart, Tasmania 7001, Australia;* [3]*Present address: Department of Zoology, University of Melbourne, Melbourne, Victoria 3010, Australia;* [4]*Present address: Ministry of Fisheries, P.O. Box 2526, Wellington, New Zealand;* **Author for correspondence (e-mail: rossdj@unimelb.edu.au; fax: +61-3-8344-7909)*

Received 28 August 2001; accepted in revised form 10 May 2002

Key words: Asterias amurensis, ecological variability, impacts, introduced species, multiple methods, predation, seastar, soft sediments, Tasmania

Abstract

Introduced species are having major impacts in terrestrial, freshwater, and marine ecosystems worldwide. Given that resources for management are limited and that only a small percentage of invaders are likely to cause large ecological change, management priorities should be based on the severity of immediate and anticipated impacts on native assemblages and commercial species. This paper synthesizes work on the current and predicted impacts of an introduced predatory seastar (*Asterias amurensis*) on soft sediment assemblages, including native species subject to commercial fishing, in the Derwent Estuary and other areas of southeast Tasmania. Due to the absence of baseline data prior to the arrival of the seastar and the presence of other anthropogenic stressors in the estuary, estimating the impact of the seastar is difficult. To help overcome the weaknesses of any single method, our assessment of impact rests on 'weight of evidence' from multiple approaches. Results from experimental manipulations at small scales, detailed observations of feeding, and field surveys over a range of spatial scales in areas with and without the seastar provide strong evidence that predation by the seastar is likely to be responsible for the decline and subsequent rarity of bivalve species that live just below or on the sediment surface in the Derwent Estuary. The data suggest that should seastar densities in other areas on the Tasmanian coast attain the current levels in the Derwent Estuary, there are likely to be large direct effects on native assemblages, particularly on populations of large surface dwelling bivalves, including several commercial species. Given the seastar's ability to exploit a broad range of food resources other than bivalves, and the functional importance of bivalves in native systems, we predict broader direct and indirect effects on native assemblages. We would be unable to reach these same conclusions from a single approach to assessing impacts. The overall picture from the combination of methods at different scales provides more information than the sum of the results of the separate lines of investigation.

Introduction

Background

Biological introductions due to human-aided translocation of species across and between continents and oceans are proving to be one of the greatest environmental and economic threats to biodiversity and ecosystem function (Lodge 1993; Vitousek et al. 1996; Cohen and Carlton 1998; Pimental et al. 2000). In marine and estuarine systems, the number of introductions continues to accumulate (Cohen and Carlton 1998; Hewitt et al. 1999; Coles et al. 1999), yet our understanding of the nature and magnitude of ecological impacts is limited to a handful of studies (Ruiz et al. 1999; Grosholz et al. 2000). In some cases, the impact on the recipient community has been catastrophic, such as the invasion of San Francisco Bay by the Asian clam *Potamocorbula amurensis* (Nichols et al. 1990) and the introduction of the comb jelly *Mnemiopsis leidyi* in the Black Sea (Shushkina and Musayeva 1990). Although prevention of invasions is the ideal outcome for management, in many cases species are well established before they are discovered. Eradication of established species in marine systems is usually not feasible, at least in the short term, particularly if they are widely established (but see Bax 1999; Willan et al. 2000; Culver and Kuris 2000). Given that only a small percentage of exotics is likely to cause large ecological change (Carlton and Geller 1993; Williamson 1996) and that resources for management are usually limited, management priorities should be based on the severity of immediate and, perhaps more importantly, anticipated impacts on native assemblages (Lodge et al. 1998; Parker et al. 1999).

The coastal waters of Australia have been the site of a number of significant introductions of exotic marine species (Pollard and Hutchings 1990a, b; Jones 1991; Furlani 1996; Hewitt et al. 1999). One of the more conspicuous introductions has been the northern Pacific seastar (*A. amurensis*), first recorded in the Derwent Estuary in southeast Tasmania (Figure 1) in 1986 (Turner 1992; Buttermore et al. 1994). The seastar is a native of the coasts of Korea, Japan, China, and Russia, but is also found in Alaska and northern Canada where some suggest it was introduced (McLoughlin and Bax 1993). It is thought to have been introduced to Tasmania as larvae in ballast water or as a fouling organism on ships from Japan, although the discovery of adult seastars in the water intake ('sea chest') of a vessel indicates that other methods of transfer are possible (Thresher 2000).

Since its arrival in Tasmania, the seastar has become the dominant benthic invertebrate predator in the Derwent Estuary (Figure 1) where it is considered a major threat to native assemblages (Buttermore et al. 1994; Johnson 1994; McLoughlin and Thresher 1994; Grannum et al. 1996). Although other invertebrate predators are present in the estuary such as the native seastar *Coscinasterias muricata*, none of them attain comparable densities or are able to maintain high densities over such large areas and long periods. Perhaps of

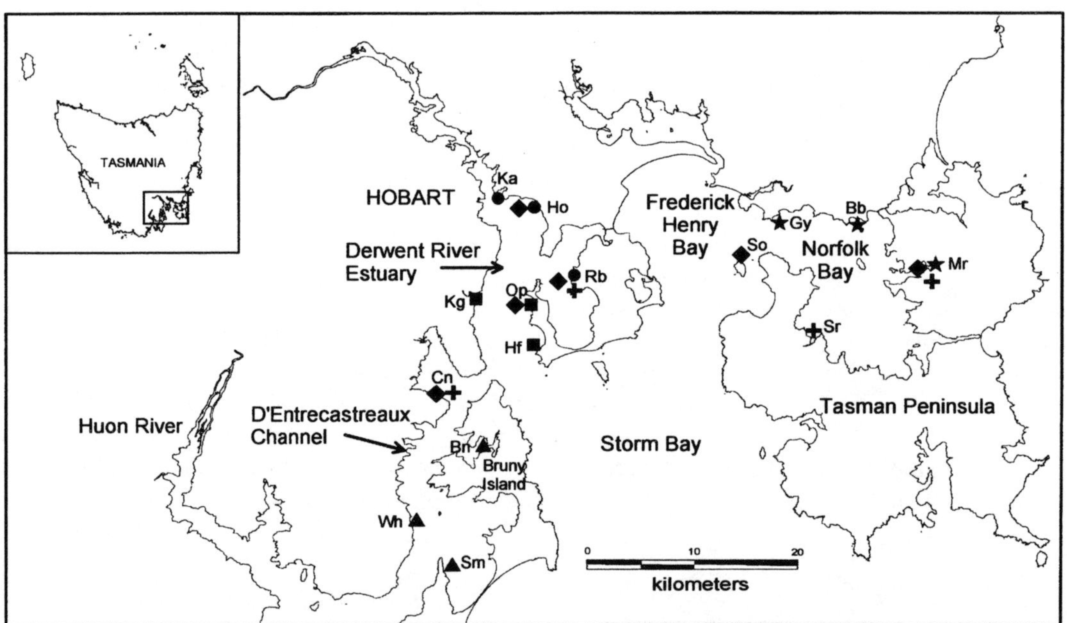

Figure 1. Map of southeast Tasmania showing experimental and survey sites in this study. Sites include those sampled in a large scale survey in 1996 (= three sites in each of four regions: two regions inside the estuary designated as upper (●) and lower (■) estuary and representing areas of high and low densities of seastars, respectively; and two regions outside the estuary where seastars were absent at Norfolk Bay (★) and in the D'Entrecasteaux Channel (▲)) and 1998 (= six sites (♦): three inside and three outside the estuary), and where manipulative experiments were conducted (+). Site codes indicate Howrah Beach (Ho), Kangaroo Bay (Ka), Ralphs Bay (Rb), Halfmoon Bay (Hf), Kingston Beach (Kg), Opossum Bay (Op), Breaknock Bay (Bb), Gypsy Bay (Gy), Murdunna (Mr), Barnes Bay (Bn), Simmonds Bay (Sm), Whaleboat Rock (Wh), Sloping Island (So), Conningham (Cn) and Saltwater River (Sr).

greater concern is the potential spread of *A. amurensis* to areas outside the estuary, locally, nationally, or internationally. Concentrations of *A. amurensis* larvae adjacent to port areas in the estuary are among the highest ever reported for seastar larvae (Bruce et al. 1995). Furthermore, recent modelling of seastar larval dispersal indicates that the majority of larvae produced in the estuary are likely to be advected from it (A. Morris and C.R. Johnson, unpub.). The recent discovery and subsequent population explosion of seastars in Port Phillip Bay, Victoria (on mainland Australia) are believed to be the result of translocation from Tasmania (Murphy and Evans 1998), and highlights the propensity of this species for range expansion. In response to the international threat posed by the seastar, New Zealand has enacted legislation preventing discharge of ballast water taken from the Derwent Estuary and Port Phillip Bay during the spawning season of the seastar (Biosecurity Act 1993, Annex 1[1]).

While there are few examples of introduced echinoderms (e.g. Ruiz et al. 1999; Cohen and Carlton 1996; Hewitt et al. 1999), the importance of asteroids in structuring benthic marine communities, their propensity for population outbreaks, and capacity to 'invade' and significantly impact fishery and mariculture grounds in their native ranges is well documented (Sloan 1980; Menge 1982). In the northern hemisphere, *A. amurensis* causes considerable damage to commercial shellfishes such as oysters, cockles, scallops, and other bivalves (Hatanaka and Kosaka 1959; Kim 1969; Nojima et al. 1986). It is also known to be an opportunistic predator on a variety of other epifaunal and infaunal species including other molluscs, ascidians, bryozoans, sponges, crustaceans, polychaetes, fish and echinoderms (Hatanaka and Kosaka 1959; Fukuyama and Oliver 1985; Fukuyama 1994). In Tasmania, indirect indications of impact from observations of seastar foraging behaviour, stomach contents, and estimates of feeding electivity suggest the potential for considerable impact on native species. Moreover, several of the native bivalve species commonly recorded in the seastars stomach (Morrice 1995; Grannum et al. 1996; Ross et al. 2002) are part of a small but growing diving-based fishery in southeast Tasmania. Nonetheless, there is no direct quantitative evidence of impacts of *A. amurensis* on native assemblages or wild fisheries in either its native or introduced range. Given that resources for management and control efforts for introduced pests in Tasmania are limited, particularly as the number of high profile invaders continues to rise (e.g. the Japanese kelp *Undaria pinnatifida*, European shore crab *Carcinus maenas* and New Zealand screw shell *Maoricolpus roseus*), management priorities should be based to as far as possible on a robust assessment of immediate and anticipated impacts on native assemblages and commercial species.

Estimating impact

A major challenge for ecologists is how to assess impacts of successful invaders in a manner that includes information on the nature, magnitude, and pattern of impact in space and time (Lodge et al. 1998; Parker et al. 1999; Ruiz et al. 1999). There are several obstacles to answering these questions. First, there are often no pre-impact data on native assemblages. In these circumstances there is no baseline for comparison, and standard robust sampling designs to assess impacts such as Before-After-Control-Impact (BACI) and its modifications (e.g. Underwood 1991, 1992) are not possible. Second, introduced species are often well established before they are discovered, and so significant impacts may have already taken place at the time of detection. Third, introduced species most often become established in areas that are subject to a broad spectrum of other anthropogenic stressors (Ruiz et al. 1999), such as in the vicinity of ports. In these locations it is difficult to separate the effects of the introduced species from other anthropogenic stressors such as pollution and physical disturbance to habitat, particularly given the likelihood of interaction between the introduced species and anthropogenic stressors (Ruiz et al. 1999). Finally, concern is usually with impacts over large spatial and temporal scales, at which manipulative experiments are difficult and normally not practical (Lodge et al. 1998; Ruiz et al. 1999).

To help surmount these difficulties, and to overcome the weaknesses of any single method of impact assessment (Diamond 1986; Schmitt and Osenberg 1996; Lodge et al. 1998; Ruiz et al. 1999), we used an integrated approach combining multiple methodologies. Direct observation of feeding and gut contents and availability of prey in the environment can provide a clear picture of prey preferences, but does not comprise either a critical test or quantitative estimate of impact. In contrast, manipulative experiments, necessarily conducted at small spatial (and usually temporal) scales, can be powerful tests of impact at particular sites. However, the generality of results of a single experiment, and whether estimates of impact from results at small spatial scales can be scaled linearly to larger scales, is far from certain. It is sound practice to repeat small scale experiments at several sites. At very least, results of small scale experiments should be consistent with large scale patterns of distribution and abundance of the introduced predator and its prey. We argue that a combination of all of these approaches provides a more robust assessment of impacts because it includes independent tests of impacts at different scales (Diamond 1986; Ruiz et al. 1999).

In this paper, we integrate our results from: (a) experiments in which seastar density is manipulated at several sites immediately beyond the current range of the seastar; (b) experiments in which seastar density is manipulated following settlement of prey at a site within the current range of the seastar; (c) experiments in which the density of both seastars and another introduced benthic predator (*C. maenas*) are manipulated and their interaction assessed; (d) comparative analysis of prey taxa in the sediments and in seastar stomachs; and (e) field surveys to examine the relationship between macrofaunal assemblages and seastar abundance at a range of spatial scales in areas with and without the seastar in southeast Tasmania. This combination of approaches provides a robust estimate of the impact of *A. amurensis* on soft sediment assemblages in southeast Tasmania.

Table 1. Summary of results from experimental manipulations at Murdunna, Saltwater River, and Conningham (sites shown in Figure 1) assessing the effects of seastar predation on the abundances of functional groups and common taxa.

	Murdunna	Saltwater river	Conningham
Surface bivalves	**0.0038** (−)	**0.0033** (−)	*
Fulvia tenuicostata	**0.0003** (−)	**0.0014** (−)	*
Katelysia rhytiphora	**0.0076** (−)	0.4863	—
Wallucina assimilis	0.1729		—
Musculus impacta	—	0.5734	—
Mysella donaciformis	—	0.0299	*
Deep bivalves	0.9120	0.5383	0.1708
Theora spp.	0.5511	0.3537	**0.0046** (+)
Laturnula rostrata	0.2920	—	
Errant polychaetes	0.5322	0.4973	0.9286
Simplisetia amphidonta	0.5862		
Glycera spp.	0.4571		
Nephtys australiensis	—	0.5529	—
Sedentary polychaetes	0.6586	0.1811	0.9357
Lysilla jennacubinae	0.4960	—	
Capitellids	—	0.2019	0.6263
Pectinaria ssp.	—	—	0.8364
Crustaceans	0.2836	0.1872	0.9871
Amphipods	0.3348	0.1030	0.7805
Ostracods	—	—	0.3086
Crabs	—	1.0000	—
Echinoderms	—	—	0.2625
Echinocardium cordatum	0.6304	0.8837	0.4125
Holothurian sp.	—	—	0.2221
Gastropods	0.5706	0.3114	0.2148

At each site there were three treatments: cage inclusion (single seastar added); cage 'control' (seastars absent); and unmanipulated plot (seastars and cage absent). Results are for the planned comparison of interest to test for the effect of seastars, i.e. cage 'control' *versus* cage inclusion. P-values <0.016 are significant for the planned comparisons (indicated in bold face). The direction of significant effects is shown in brackets; — = taxon absent from this site; * = tests not undertaken due to a clear treatment by block interaction. Table adapted from Ross et al. (2003).

Variability in impact

The precision to which the impact of introduced species can be predicted will depend largely on the magnitude of spatial and temporal variability in impacts. However, relatively few studies of introduced marine species have investigated spatial and temporal variability of impacts (but see Allmon and Sebens 1988; Nichols et al. 1990; Grosholz and Ruiz 1996), despite that research on the ecological impacts of native species suggests several factors that may influence both the nature and magnitude of effects of introduced species. These include the density of both the impacting species and its potential prey, the nature of functional responses to prey density, water temperature, current velocity, turbidity, and sediment characteristics (e.g. Lipcius and Hines 1986; Woodin 1978; Everett and Ruiz 1993; Skilleter 1994; Thrush 1999). For example, we might expect an increase in the variability of impacts in space and time if prey populations are spatially and temporally variable or if the exotic is a generalist predator with distinct prey preferences. The results of our large scale survey conducted in 1996 (see details later and in Ross 2001) highlight the magnitude of variability at a range of spatial scales (from metres to tens of kilometres) in the structure of soft sediment assemblages, sediment characteristics, and seastar density. Results of temporal surveys conducted at a subset of the same sites were also indicative of significant intra- and inter-annual variability in benthic assemblages and seastar density (Ross 2001). Based on this evidence alone we might expect the impact of the seastar to vary in space and time.

To assess the variability, and hence predictability, of the impact of the seastar on soft sediment assemblages we used experiments to examine the impact of the seastar in three sheltered bays (separated by ca. 10–45 km) immediately beyond the current range of the seastar (Ross et al. 2003). The experiments were run for 8 weeks at each site and included three treatments: a cage inclusion (single seastar added), cage 'control' (seastars absent), and an unmanipulated plot (seastars absent, no cage) set up as a randomized complete block design, i.e., each of the three treatments (≈5 m apart) were applied to each of three blocks (≈30 m apart). This design was necessary given large spatial variability in the soft sediment assemblages at scales of 30–40 m, and practical limitations to the total number of experimental plots that could be maintained and sampled. The experimental plots were 1 m^2, and thus, seastar density was 1 m^{-2} in the cage inclusion. This is less than the maximum densities of the seastar in the Derwent Estuary (Grannum et al. 1996; S Ling unpublished data). Cages consisted of a rigid (1 m × 1 m base × 0.7 m high) steel frame with protruding legs (0.5 m long). The cage top and sides were covered in stiff plastic mesh (6 mm), and the cage legs driven into the sediment so that 100–150 mm of the cage sides was buried to prevent passage of large predators or prey by burrowing.

Given that the experiments were conducted at different locations and at different times (because it was logistically too difficult to run them simultaneously), it is perhaps not surprising to see marked differences in the soft sediment assemblages among sites at both the species and functional group level (Figure 2). Consequently, seastar impacts were not qualitatively equivalent across sites (Ross et al. 2003). Nonetheless, when potential prey taxa were separated into functional groups that also reflected their likely ecological availability to the seastar (e.g. surface dwelling *versus* deep burrowing bivalves), a consistent negative effect of the seastar on surface dwelling bivalves at all sites was clearly evident (Figure 3). Bivalves are known to be a major food source of the seastar in both its introduced (Grannum et al. 1996; Lockhart and Ritz 2001b) and native ranges (Hatanaka and Kosaka 1959; Nojima et al. 1986). Differential impacts on surface-living taxa and deeply burrowing taxa have similarly been demonstrated for other predators in soft sediment habitats (e.g. Woodin 1974; Virnstein 1977).

While the effect of seastar predation on surface bivalves was common across all sites, the magnitude of the impact varied between sites (Figure 3). Seastar density at 1 m^{-2} produced decreases of 92 m^{-2} (96%), 35 m^{-2} (72%), and 31 m^{-2} (40%) at Murdunna, Saltwater River, and Conningham, respectively. This pattern reflected the relative abundance of large surface bivalves (>8 mm) at each site (92%, 67%, and 41% of total bivalves respectively; Figure 3). Although there was a small decrease in abundance of the 2 smaller size classes (2–4 and 4–8 mm) in the presence of seastars, the change in abundance of larger bivalves was largely responsible for the overall decrease in bivalves in the presence of seastars. These results are consistent with the patterns evident in a large scale survey we conducted in 1998 (see details later and in Ross 2001) comparing the abundance of bivalves and heart urchins at sites with and without seastars (Figure 4; see later). Larger bivalves (>8 and 4–8 mm) were extremely rare at sites where seastars were abundant relative to sites without seastars (Figure 4c). However, it is unclear whether the observed patterns in the experiments and survey are a consequence of size or species selection (or both). In these experiments we cannot differentiate between size and species selection because the species that were most heavily impacted at each site were also the large (>8 mm) species of surface bivalves. However, in the survey, bivalve species identified as important prey (on the basis of results in the small scale experiments and feeding observations) were rarer in the presence of seastars irrespective of their size, which suggests that species-level selection operates. Given that laboratory and field observations have demonstrated both size and species selection by *A. amurensis* (Lockhart and Ritz 2001a, b), changes in both size and species composition may underpin variability in the magnitude of impact between sites.

At Conningham the results indicated variability in impact at even smaller spatial scales, i.e., among blocks separated by ca. 30 m (Ross et al. 2003). In contrast to other sites, epifaunal bivalves, predominately *Electroma georgina*, were common at Conningham. Reduced densities of *E. georgina* but not surface dwelling infaunal bivalves indicated predation by the seastar on *E. georgina* in two of the three blocks. In one block, however, where *E. georgina* was rare to begin with, the density of surface dwelling infaunal species declined. We interpret this pattern to reflect both the ecological availability of prey and small scale patchiness of *E. georgina*. In patches where *E. georgina* is abundant, since it is epifaunal and therefore, directly accessible to the seastar, the seastar apparently preferentially consumes *E. georgina*. In patches when *E. georgina* is rare or absent, the seastar will readily consume infaunal species of surface dwelling bivalves.

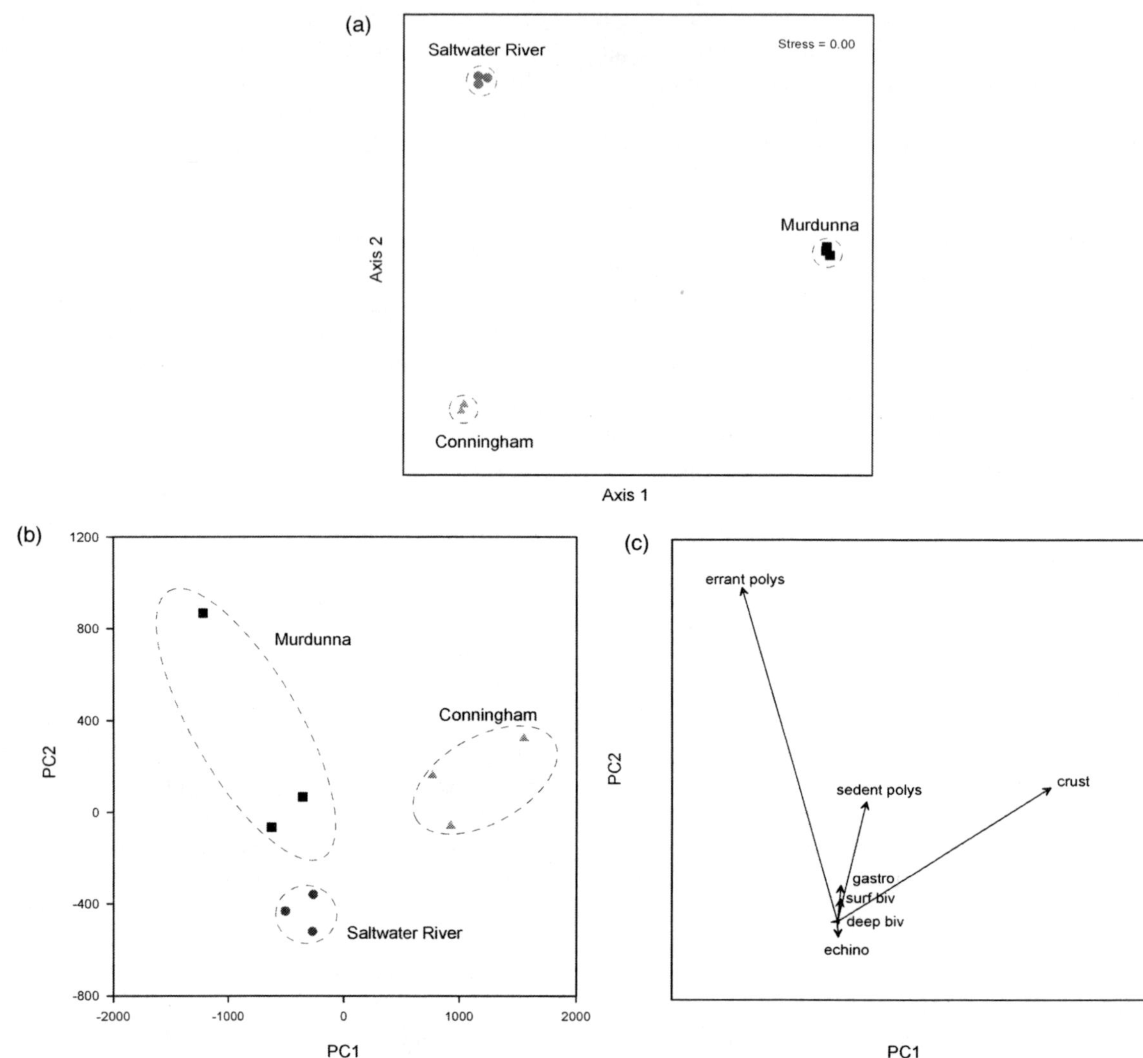

Figure 2. Composition of soft sediment assemblages at the three experimental sites where seastar effects were assessed. The sites are separated by 10–45 km (sites shown in Figure 1). (a) MDS ordination of three replicate unmanipulated plots at each site based the Bray Curtis matrix of 4th root transformations of densities of macro-benthic species, and (b) a PCA plot of functional groups across the same locations and plots, reveal distinct differences among sites in the composition of assemblages (principal components 1 and 2 account for 94% of the total variance). (c) The biplot associated with the PCA identifies the functional groups most responsible for the patterns shown in the PCA plot (crust = crustaceans, errant polys = errant polychaetes, sedent polys = sedentary polychaetes, surf biv = surface dwelling bivalves, deep biv = deep burrowing bivalves, gastro = gastropods, and echino = echinoderms). Figure adapted from Ross et al. (2003).

In the nearby Derwent Estuary, feeding surveys in which we compared the relative abundance of prey in seastar stomachs relative to their occurrence in the environment (see Ross 2001 for more details) highlight the likelihood that temporal changes in the composition of assemblages also generate variability in impacts. At Opossum Bay, the surface infaunal bivalves *Timoclea cardoides, Venerupis anomala*, and *F. tenuicostata* were major prey items of the seastar except when *E. georgina* became available at high densities, at which time the epifaunal bivalve dominated the seastar's diet (Ross, unpub. data). Similarly, at Ralphs Bay the seastar fed predominately on bivalves, but fed largely on other species (e.g. crustaceans and the heart urchin *E. cordatum*) when bivalves became relatively rare (Ross et al. 2002; Figure 5).

The generalist nature of feeding in asteroids is well known (e.g. Menge 1982). Dietary composition

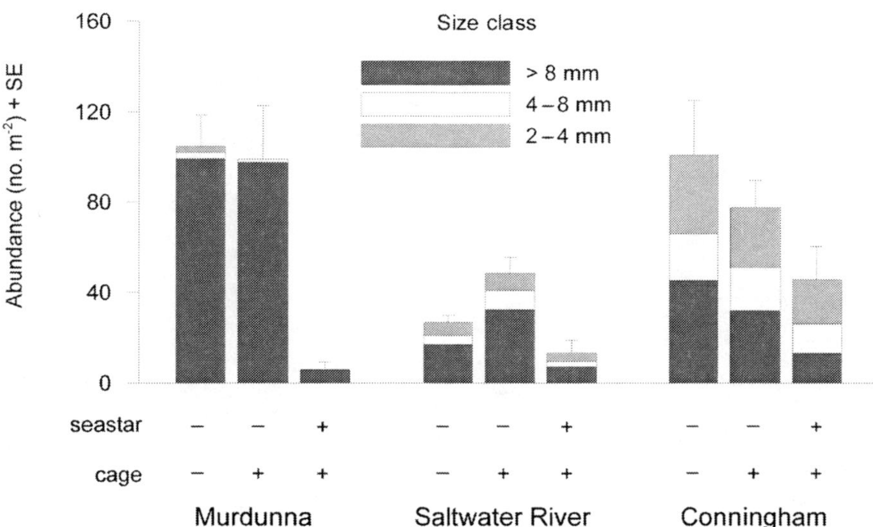

Figure 3. Results from experimental manipulations at Murdunna, Saltwater River, and Conningham (sites shown in Figure 1) assessing the effect of seastar predation on surface dwelling bivalves. At each site there were three treatments: cage inclusion (single seastar added); cage 'control' (seastars absent); and unmanipulated plot (seastars and cage absent). Density is the mean per 1 m^2 (+SE). The proportion of total bivalves retained on 2, 4, and 8 mm mesh are also depicted. Figure adapted from Ross (2003).

often tracks changes in the relative availability of prey species. Our results demonstrate that the exact nature of the effect of seastar predation on soft sediment assemblages is likely to be site- and time-specific depending on spatial and temporal variability of prey species at the time of arrival of the seastar at a site, and the length of time that seastar populations have been established at particular sites. Preferred prey (large surface dwelling bivalves) are consumed when the seastar first establishes at a site, but as preferred items become rare, the seastar switches to other species.

Impact on the survivorship of juvenile prey

Despite the presence of numerous remains (intact shells) of large adult bivalves in the surface sediments (Lockhart 1995; Ross, pers. obs.), live bivalves >5–10 mm are now rare in the Derwent Estuary where seastars are abundant. The results of the experiments summarized above are consistent with the notion that predation by the seastar is responsible for this rarity, particularly of shallow infaunal and epifaunal species. The high prevalence of juvenile molluscs, and particularly bivalves, in the diet of the seastar suggests that seastar predation may potentially also be preventing re-establishment and recovery of adult populations in the estuary.

In early 1998, a massive settlement of the commercial bivalve *F. tenuicostata* was recorded at Ralphs Bay (Figure 1) in the Derwent Estuary, where seastars are abundant. This provided an ideal opportunity to test whether seastar predation might be limiting the survivorship of *F. tenuicostata* juveniles, and therefore, the subsequent recovery of adult populations. The effect of seastar predation on the survival of *F. tenuicostata* juveniles was examined in a caging experiment conducted over a 10-week period. Cages were identical to those described previously, and this experiment was also established as a randomized complete block, but with different suite of treatments to that outlined earlier. The treatments included an unmanipulated plot subject to background predation and sampled at the start of the experiment, an unmanipulated plot subject to background predation and sampled at the end of the experiment, a caged inclusion (one seastar added) and a caged exclusion (no seastars present). Further details are in Ross et al. (2002). There was a ≈9-fold reduction in the density of *F. tenuicostata* juveniles (from 300 to 35 m^{-2}) in the presence of seastars in unmanipulated plots over the experimental period (O_{initial} vs. O_{final}; Figure 6a), and a ≈34-fold reduction in the density of *F. tenuicostata* juveniles (from 580 to 17 m^{-2}) in the presence of seastars in caged exclusions relative to caged inclusions ($-S$ vs. $+S$; Figure 6a). Note the increase in density over

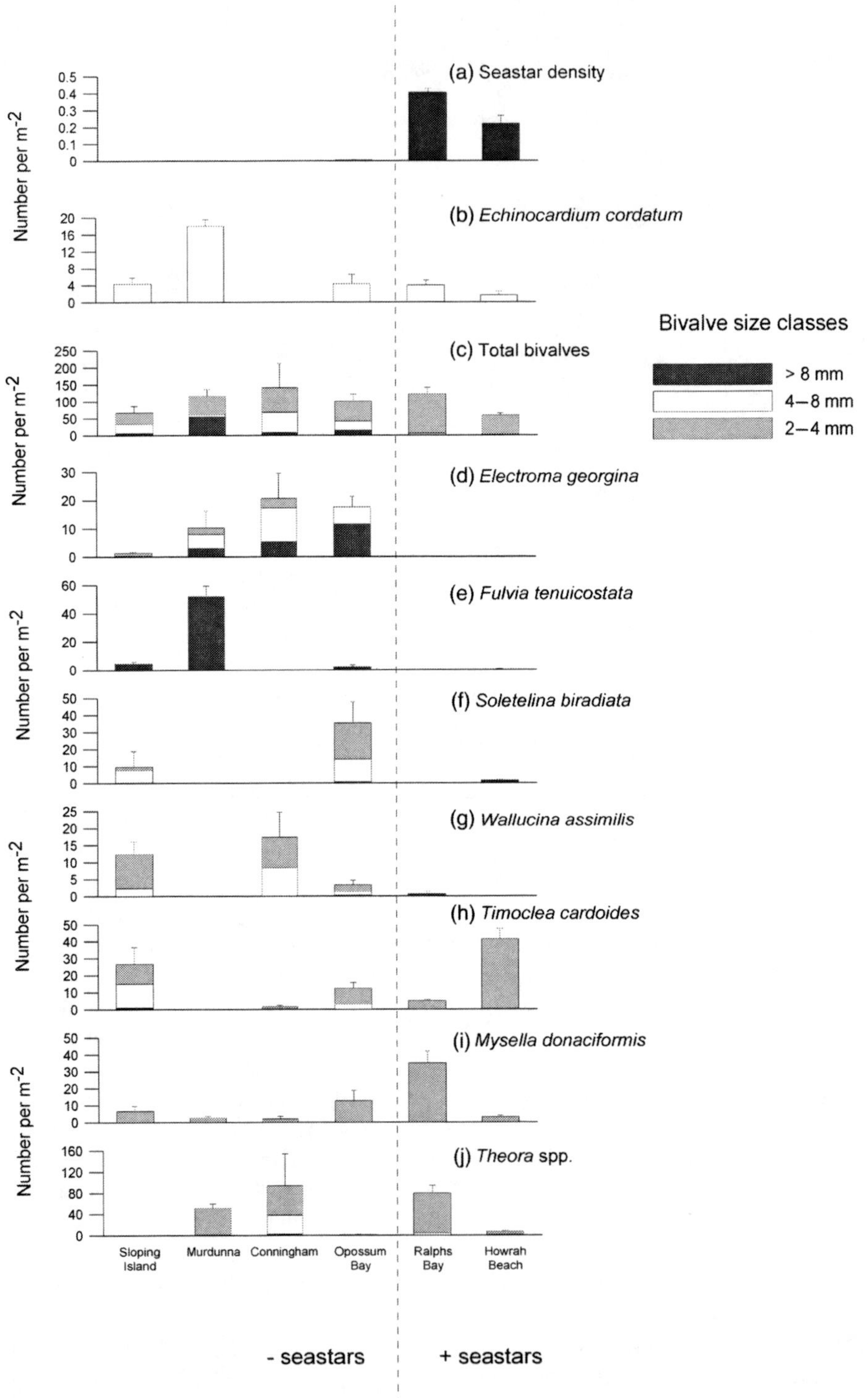

Figure 4. Results from the large scale survey in 1998 that used suction samples (1 m^2, $n = 3$) to estimate the abundance of bivalves and heart urchins at sites with and without seastars (design as in Figure 1). Mean densities (+SE) of (a) seastars based on 50 m × 2 m strip transects ($n = 3$); (b) *E. cordatum* (heart urchin); (c) total bivalves; and (d–j) each of the major bivalve species at each site. The proportion of bivalves retained on 2, 4, and 8 mm mesh are depicted. Figure adapted from Ross 2001.

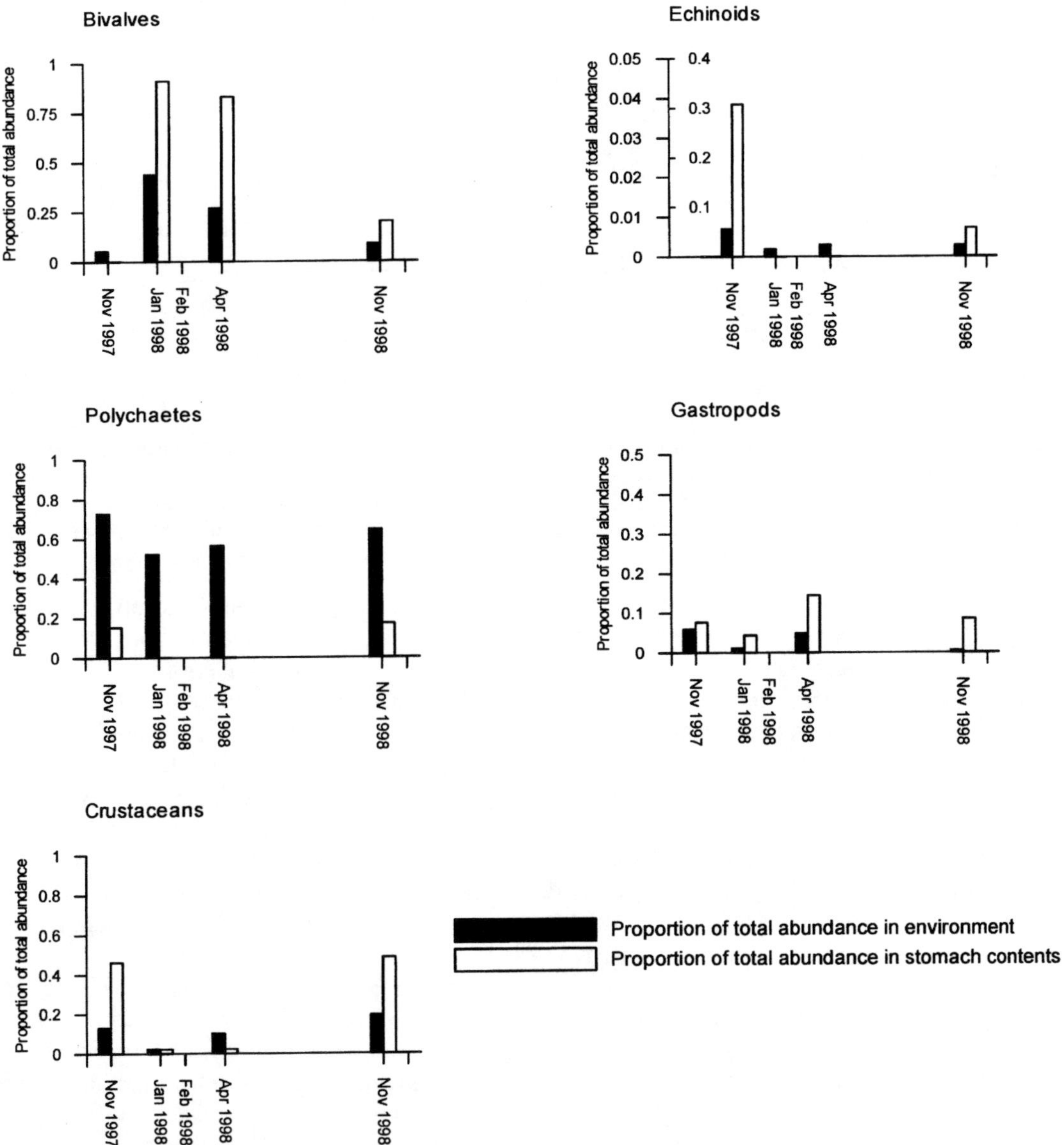

Figure 5. Proportion of the total abundance of the major prey groups in the diet and sediments at Ralphs Bay (Note: when two scales are marked, the left-hand scale is the proportion of total abundance in the sediments.) Figure adapted from Ross et al. (2002).

the experimental period (i.e. from $300\,\mathrm{m}^{-2}$ $\{O_{\mathrm{initial}}\}$ to $580\,\mathrm{m}^{-2}$ $\{-S\}$) most likely reflects the growth of juveniles into size classes that are retained on the 2-mm mesh sieve that was used to assess abundances. Densities of ca. $530\,\mathrm{m}^{-2}$ were recorded at the start of the experiment when juveniles retained on a 1-mm mesh sieve were counted.

To identify whether there was a shift in diet of *A. amurensis* concomitant with the settlement of *F. tenuicostata*, we conducted a feeding survey to compare the relative abundance of prey in seastar stomachs relative to their occurrence in the environment before, during, and after the settlement event (Ross et al. 2002). The survey showed clearly that the seastar responded to the settlement of *F. tenuicostata* with a pronounced shift in diet. The bivalve was the most common prey species of the seastar following its settlement, representing 50–80% of the dietary items in February and April 1998 (Figure 6b), while the seastar later shifted diet to feed on other species when the

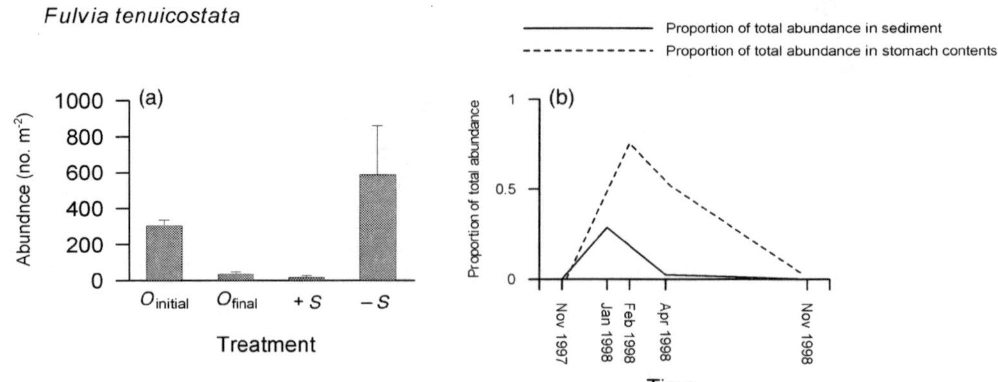

Figure 6. Results of the experimental manipulation and feeding survey following settlement of the commercial bivalve *F. tenuicostata* at Ralphs Bay. (a) Density of *F. tenuicostata* juveniles in four treatments over the 10-week experimental period. Treatments are unmanipulated plots sampled at the start of the experiment soon after the settlement event (O_{initial}); unmanipulated plots sampled at the end of the experiment, subject to background levels of predation (O_{final}); caged inclusion sampled at the end of the experiment with a single seastar added to 1 m^2 cages ($+S$); caged exclusion sampled at the end of the experiment, no seastars present ($-S$). Density is the mean per 1 m^2 (+SE, $n = 3$ plots). (b) Proportion of the total abundance of *F. tenuicostata* in seastar stomachs and in sediments. Figure adapted from Ross et al. (2002).

bivalve became relatively rare. Overall, the population of this bivalve was virtually eliminated within 10 weeks of settlement. The feeding observations following the decline of *F. tenuicostata* indicated that seastar predation might also be limiting the survivorship of juveniles of an introduced bivalve, *Corbula gibba*. *C. gibba* had not been previously reported from the Derwent Estuary, but is present in high densities in the nearby D'Entrecasteaux Channel where seastars are comparatively rare. In Port Phillip Bay on mainland Australia, *C. gibba* is considered a major prey item of *A. amurensis* (G. Parry pers. comm.). Thus, it is plausible that seastar predation on *C. gibba* juveniles may be impeding the establishment of the bivalve in the Derwent Estuary. The effect of seastar predation on the survivorship of juvenile bivalves is also evident in the mariculture industry, particularly in the collection of scallop spat, where losses of commercial spat due to *A. amurensis* over a settlement season are reported to be as high as 50% (S. Crawford, pers. comm.).

Interactions with other factors

Interactions between introduced species and other anthropogenic stressors may greatly influence the impact of introduced species (Ruiz et al. 1999). Estuaries and bays, which represent the most invaded habitats in coastal regions, are also often the most degraded coastal habitats. The Derwent Estuary is no exception (Coughanowr 1997; Bennett 1999). A common generalization is that disturbed habitats are more readily invaded than pristine ones, largely because of reduced competition or predation (Elton 1958; Lodge 1993; Ruiz et al. 1999; Simberloff and Von Holle 1999). Given the degraded state of the Derwent Estuary, lack of native predators and/or competitors of the seastar may have played a major role in the successful establishment of the seastar in the estuary (Bennett 1999), although the idea has not been tested experimentally. Regardless, the seastar is clearly the most numerically dominant benthic invertebrate predator in the soft sediment assemblages of the Derwent Estuary.

Recently, seastars have been increasingly recorded in more 'pristine' areas outside the Derwent Estuary, and interaction with other predators and competitors appears inevitable. Ironically, potential spatial overlap and interaction with another introduced species in Tasmania, the European green crab (*C. maenas*) is anticipated since both species are major predators of bivalves in sheltered low energy environments (Griffiths et al. 1992; Grosholz and Ruiz 1995). It is often the case that effects of multiple species cannot be predicted from estimates of the effect of each species alone due to complex interactions (Kareiva 1994). The question of the combined effects of several introduced species is particularly germane given the possibility that synergistic effects may lead to accelerated impacts on ecosystems with the addition of each new species (Simberloff and Von Holle 1999).

We examined the separate and combined impacts of *Asterias* and *Carcinus* in a manipulative experiment that ran for eight weeks (Ross et al., in press). The seastar/crab/bivalve interaction is a useful model to test for interaction modifications between predators because the different predators leave a characteristic trace of their feeding activities. Undamaged, empty shells identify bivalves eaten by seastars, whereas bivalve hinges with only a fraction of the shell remaining identify bivalves eaten by crabs. The five treatments used to investigate predation effects and test for potential cage effects included all possible combinations of presence (a single animal per cage) and absence of crabs and seastars in cages, and an unmanipulated 1 m^2 plot without either cages or added predators. Cages were identical to those described previously, and this experiment was also established as a randomized complete block, i.e., each of the five treatments ($\approx$5 m apart) was applied to each of three blocks ($\approx$30 m apart). Each predator had a major effect on the abundance of bivalves, particularly the commercial bivalve *F. tenuicostata* (Figure 7). However, when both predators were present simultaneously their impact on *F. tenuicostata* was different to when they were alone. The number of *F. tenuicostata* hinges (indicative of predation by *Carcinus*) in the presence of both predators was reduced compared to when the crab was alone, but higher than when these predators were absent (Figure 7b). While the number of open shells (indicative of predation by *Asterias*) in the presence of both predators was similar to when the seastar was alone (Figure 7b), fewer small bivalves were consumed compared to when the crab was absent (Figure 8). The interaction between *A. amurensis* and *C. maenas* is likely to be one of resource competition, resulting in some partitioning of bivalves according to size, with *A. amurensis* consuming the large and *C. maenas* the smaller individuals (Figure 8).

While there are limitations in extrapolating to larger spatial scales on the bases of results of experiments conducted at small scales (e.g. Thrush et al. 1997, 1999; Lodge et al. 1998; Fernandes et al. 1999), there are nonetheless some predictions on the impact of *C. maenas* and *A. amurensis* on a larger scale that can be made. The zone of overlap of these two species is likely to be largely restricted to the shallow subtidal given that *C. maenas* occurs predominately in the intertidal through to the shallow subtidal while *A. amurensis* occurs predominately in the shallow to deep subtidal. If the predators coexist in the shallow subtidal the results suggest that the effect of each predator is likely to be modified in the presence of the other should they attain the densities used in this experiment. Although the densities in this experiment are high, they are well within the range observed for both species in their respective ranges in Tasmania, particularly where bivalves are abundant. On a larger spatial scale, the combined effect on bivalve populations may be greater than that due to each predator alone, simply because their combined distribution covers a broader range of habitats. Overall, we predict that the two introduced predators are likely to coexist because of resource partitioning according to size and/or different habitat preferences.

Large scale patterns

Although manipulative experiments are powerful tools for identifying impacts, they are typically limited to

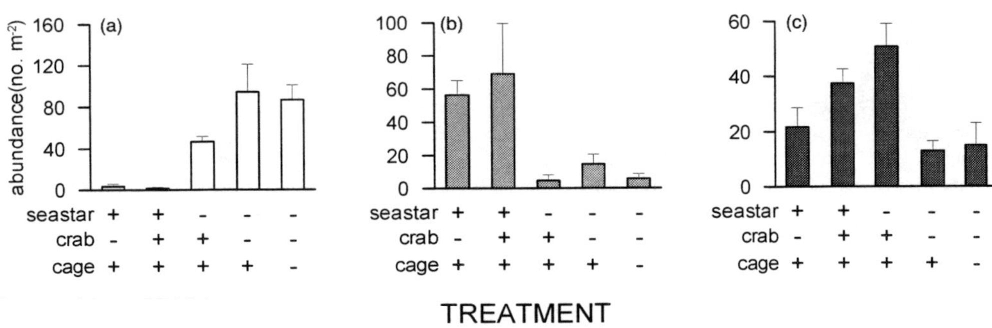

Figure 7. Results of an experimental manipulation conducted at Murdunna (site shown in Figure 1) to assess the interaction of *A. amurensis* and *C. maenas* in their effect on the commercial bivalve *F. tenuicostata*. Density of (a) alive animals, (b) open shells (indicative of seastar predation), and (c) hinges (indicative of crab predation) in each treatment. The five treatments used to investigate predation effects and test for potential cage effects included all possible combinations of presence (a single animal per cage) and absence of crabs and seastars in cages, and an unmanipulated plot without either cages or added predators. Density is the mean per 1 m^2 (+SE, n = 3 plots). Figure adapted from Ross et al. (in press).

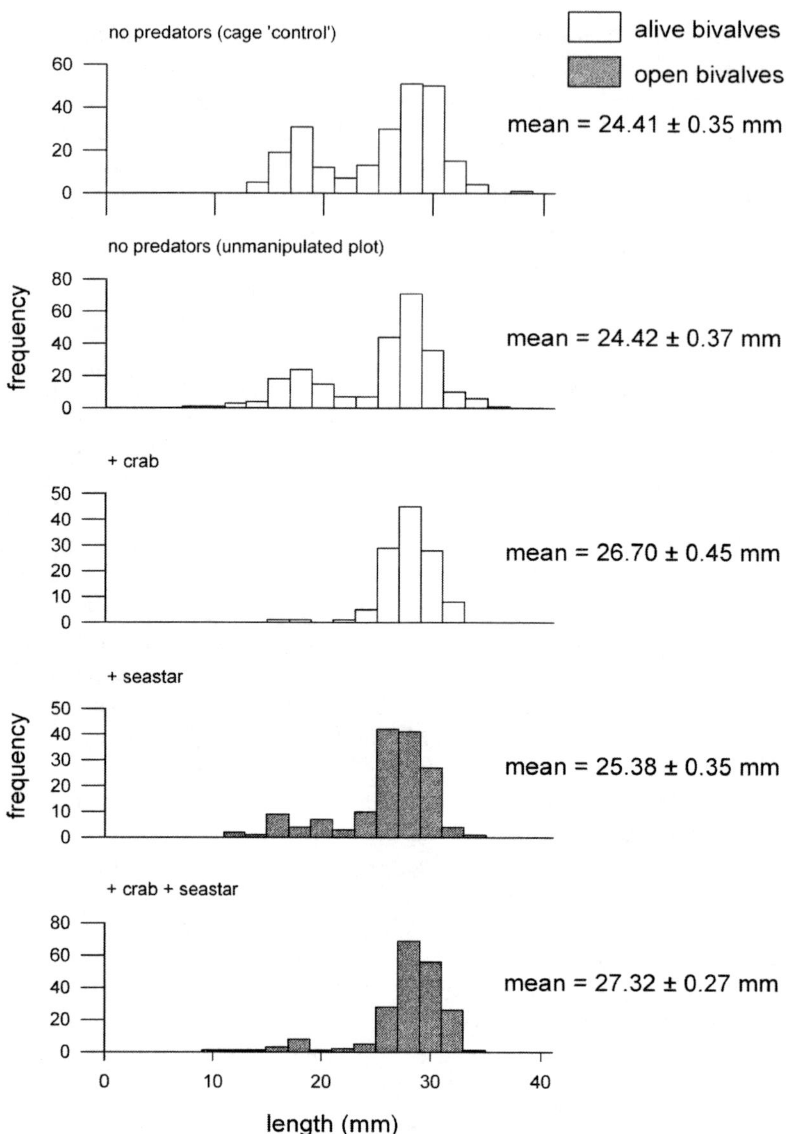

Figure 8. Length frequency histograms of open (treatments with an added seastar, and seastar + crab) and live (treatments with an added crab, cage 'control' with no added predators, and unmanipulated plot with no predators added) *F. tenuicostata* remaining at the end of experimental manipulations at Murdunna to assess the interaction of *A. amurensis* and *C. maenas* in their effect on native species. The five treatments used to investigate predation effects and test for potential cage effects included all possible combinations of presence (a single animal per cage) and absence of crabs and seastars in cages, and an unmanipulated 1 m^2 plot without either cages or added predators. Figure adapted from Ross et al. (in press).

small scales in space and time (Underwood 1996; Thrush et al. 1997; Lodge et al. 1998). Given that managers are often concerned with impacts over large spatial and temporal scales, large scale surveys are widely used as a basic tool in field assessments of environmental impacts (Osenberg and Schmitt 1996). In 1996 we conducted a large scale spatial survey comparing the composition of soft sediment assemblages at locations with ('impact' sites) and without seastars ('control' sites), and between areas with high and low seastar densities (Ross 2001). The survey included three sites in each of four regions: two regions inside the estuary designated as upper and lower estuary and representing areas of high and low densities of seastars respectively, and two regions outside the estuary where seastars were absent. At each site, there were three plots approximately 30 m apart. In each plot, two replicate benthic samples approximately 1 m apart were collected with a corer (150 mm diam., 100 mm depth). Thus, the design was hierarchical, incorporating five spatial scales of sampling (details of the survey are presented in their entirety in Ross 2001). The structure of soft sediment assemblages was highly variable at a range of spatial scales from metres to tens of kilometres. Clear differences in the multivariate composition of assemblages (Figure 9a) and abundances of some of the major taxa (Figure 10) were detected between areas with and without seastars and between areas with low and high seastar densities. However, the observed differences are more likely due to differences in sediment characteristics rather than reflecting impacts of the seastar (Figures 9 and 10). For instance, sites with seastars present were generally correlated with relatively low levels of silt but a higher proportion of fine sands (Figures 9b, c). Using the BIOENV procedure in the PRIMER programme we identified the combination of nine 'environmental' variables assessed (eight sediment characteristics and seastar density) that best correlated with patterns in the biotic similarity matrix (Clarke and Ainsworth 1993). While the combination of percentage of silt, fine sand, medium sand, and seastar density provided the best correlation with the composition of faunal assemblages ($r_s = 0.463$), when variables were examined on their own, seastar density showed the poorest correlation with the composition of assemblages (silt $r_s = 0.165$, fine sand $r_s = 0.400$, medium sand $r_s = 0.175$, and seastar density $r_s = 0.032$). The importance of sediment characteristics as a correlate and/or determinant of patchiness in soft sediment assemblages is well known (e.g. Sanders 1958; Rhoads 1974; Barry and Dayton 1991; Hughes et al. 1972). In this study the association between sediment characteristics and the composition of assemblages appears to be largely attributable to polychaetes (Ross 2001). The greatest abundances of errant polychaetes were in poorly sorted sediments with a low percentage of (<30%) fine sands (Figures 10a, f). Given the numerical dominance of polychaetes in our benthic samples and their strong association with sediment characteristics, it is not surprising that the composition of the assemblage as a whole correlated with sediment characteristics. Since our experiments showed no evidence that polychaetes are affected by seastar predation, we suggest that seastar abundance is likely to play a minor role relative to sediment characteristics in influencing the broad patterns in the total composition of assemblages that we observed (Ross 2001). Notably, if we had interpreted these results on the basis of the abundance of seastars alone, we may have erroneously inferred that the presence or absence of the seastar was responsible for the distinct differences in assemblages inside and outside the Derwent Estuary.

Because we did not expect that core samples would provide accurate estimates of the abundance of larger and/or rare species (or even necessarily detect them) that may be important prey species, we conducted a second survey in 1998 using much larger samples (1 m^2) but across a more limited spatial extent (Ross 2001). In light of the findings of feeding observations and experimental studies we focused on species that were identified as preferred prey items (bivalves and heart urchins). Small differences in sediment characteristics detected across sites did not explain the large scale patterns of abundance and size structure of prey species (Figure 11). Large scale patterns of abundance and size structure were consistent with seastar effects anticipated from work at small scales for some, but not all species (Figure 4). For example, species that were most heavily predated or preferred by seastars in experiments and/or feeding observations (i.e. *F. tenuicostata*, *E. georgina*, *Soletellina biradiata*, and *Wallucina assimilis*) were abundant only at sites where seastars were absent (Figures 4d–g). Conversely bivalves that are found rarely in seastar stomachs (*Mysella donaciformis* and *Theora* spp.) were abundant at some sites irrespective of seastar abundance (Figures 4i, j). Two species that we expected to be rare in the presence of seastars based on known feeding preferences (Grannum et al. 1996; Lockhart and Ritz 2001b; Ross et al. 2002) were the heart urchin *E. cordatum* and the bivalve *Timoclea cardoides*, however both species were abundant at some sites with and without seastars at high densities (Figures 4b, h). For *E. cordatum*, this may be because it has a depth refuge from seastar predation due to its ability to remain deeply buried (up to 150 mm: Buchanan 1966). Although the pattern of total abundance for *T. cardoides* was also not indicative of effects of seastar predation, differences in the size structure were suggestive of effects of the seastar since individuals of this species were smaller at sites were seastars were present than at sites where they were absent (Figure 4h; Ross 2001).

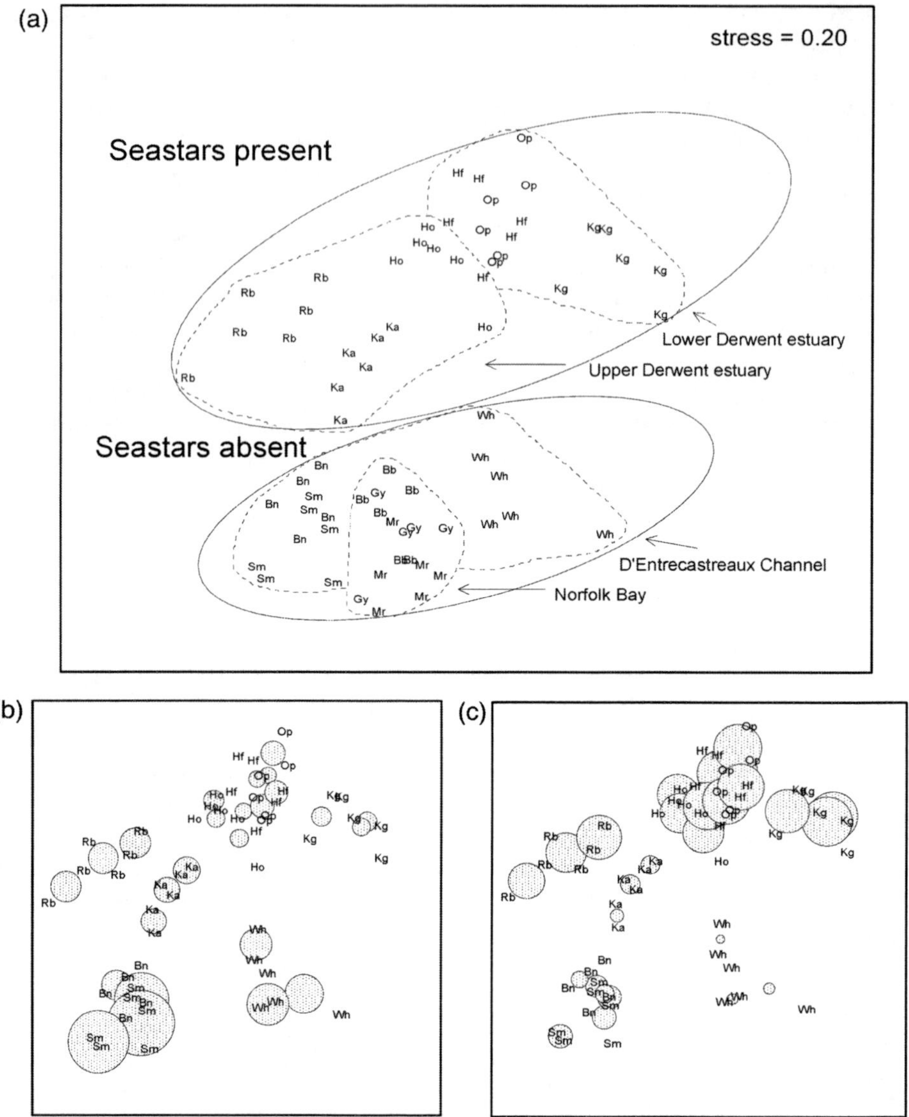

Figure 9. (a) MDS ordination based on Bray Curtis similarity matrix derived from 4th root transformation of taxon densities for all core samples in the large scale survey in 1996 (design and abbreviations as in Figure 1). Samples from zones with and without seastars, and from regions within zones are identified. Bubble plot overlays of the MDS ordination show percentages of (b) silt and (c) fine sands at each site (note that sediment data were not available for the Norfolk Bay sites). Figure adapted from Ross (2001).

Multiple methods of impact assessment

There is little doubt that large scale monitoring programmes can provide evidence of impacts. However, when there are no pre-impact data, as is often the case with introduced species, inferences about potential cause–effect relationships rely on correlations based on differences between putative 'impact' and 'control' locations. This is problematic because it is impossible to determine whether impact sites have not always been different, because there is no evidence that any change has occurred (e.g. Keough and Mapstone 1995). While this highlights the importance of identifying and accounting for environmental factors that may generate spatial variability when choosing control sites (Glasby 1997; Keough and Mapstone 1997; Glasby and Underwood 1998), the results of our large scale survey demonstrate the difficulty in selecting appropriate

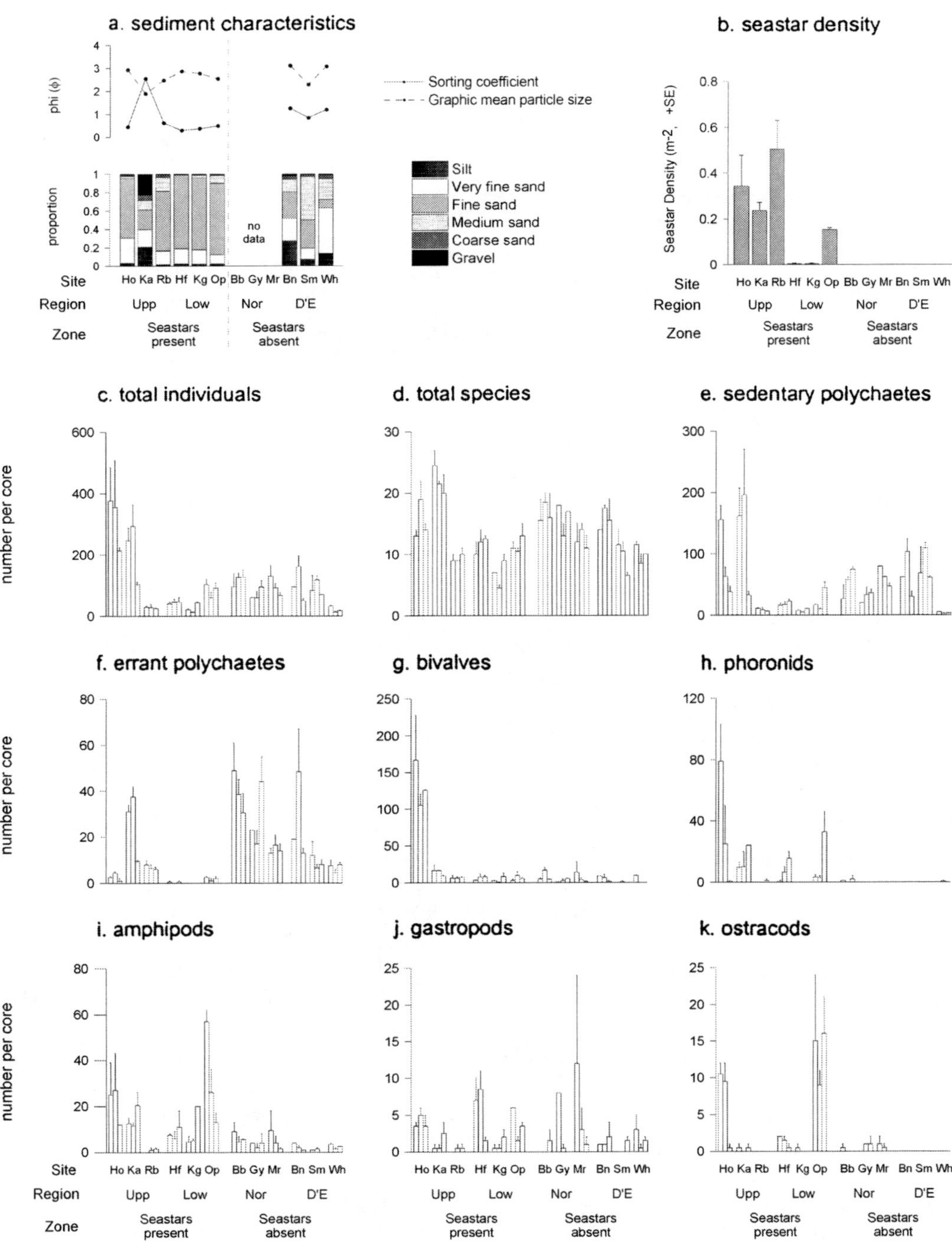

Figure 10. Summary of results from the large scale spatial survey undertaken in 1996 (design and abbreviations as in Figure 1). (a) Sediment characteristics at each site (determined from three sediment cores; note there were no data available for the Norfolk Bay sites); (b) seastar density at each site (densities are means (+SE) determined from three 50 m × 2 m strip transects); (c) mean abundance of the total number of individuals of all species; (d) total number of species; and (e–k) the major groups (data are the means (+SE) of two replicate cores taken from each plot (n = 3) at each site). Figure adapted from Ross (2001).

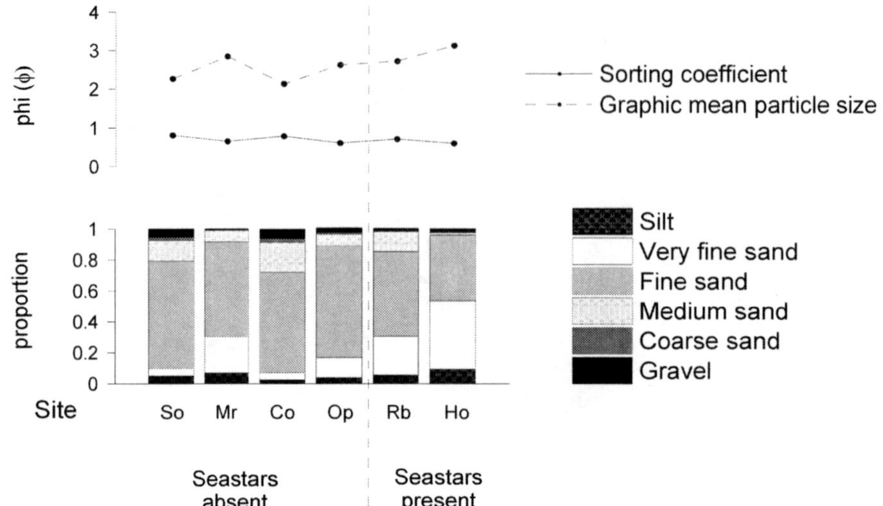

Figure 11. Sediment characteristics at each site (determined from three sediment cores) sampled in the large scale spatial survey in 1998 (design and abbreviations as in Figure 1).

control locations. Despite our efforts to use sheltered bays with similar sediment characteristics, superficial classification of sediment types proved inadequate. We detected clear differences in faunal assemblages at 'impact' sites (with seastars) and putative 'control' sites (without seastars; Figure 9a), but these differences are more likely attributable to sediment characteristics than to seastars. Properly conceived experiments do not have the inherent limitations of large scale surveys when pre-impact data are lacking. We have shown that small scale manipulative experiments conducted over relatively short periods and at the boundaries of the distribution of the introduced species, can clearly identify the nature of potential impacts. However, while small scale experiments can be extremely powerful at assessing cause–effect relationships it cannot be assumed that quantitative estimates of effect sizes measured at small scales in space and time necessarily 'scale up' (see Diamond 1986; Thrush et al. 1997, 2000; Lodge et al. 1998). Our work has suggested that, at very least, small scale experiments should be repeated at several sites to assess variability in the response to particular manipulations. We found quantitative and qualitative differences both among and within sites in responses to identical treatments, largely reflecting variation in the composition of native assemblages, the preferences of a generalist predator, and the capacity of the predator to rapidly alter its dietary composition. Notwithstanding these differences, the experiments and feeding observations showed that seastars have a consistent impact at a functional group level, namely in their impact on relatively large epifaunal or shallow infaunal bivalves. These results are consistent with patterns observed in our large scale surveys.

Overall, in assessing the impact of an introduced species in the absence of data prior to its establishment, we suggest that our study highlights the utility of employing a 'weight of evidence' approach from multiple lines of evidence. The combination of experimental manipulations and feeding observations proved crucial in identifying key ecological variables most likely to be affected by seastars. This assisted in both the design and interpretation of the large scale surveys which were ultimately consistent with impacts on some species (e.g. large bivalves) but not others (e.g. small bivalves and polychaetes).

Conclusions and implications for management

There is strong evidence that predation by *Asterias* is responsible for the present rarity of bivalve species that live just below or on the sediment surface in the Derwent Estuary. Anticipated impacts on native assemblages, wild fisheries and mariculture operations in areas outside the Derwent Estuary are of immediate management concern. We predict that should the seastar attain high densities in areas outside the estuary, there are likely to be large effects on native assemblages, particularly on large bivalves that live just under or on the sediment surface. Furthermore, since spatial overlap with the introduced green crab *C. maenas* appears imminent as their respective ranges expand, impacts on bivalve populations may be even greater than have been observed in the estuary. This is a particular concern to the small but growing number of commercial operations harvesting wild populations of the relatively large bivalves that live near the sediment surface such as *F. tenuicostata*, *Katelysia* spp., *Venerupis* spp., *Bassina* sp., *Mactra* sp., and *Spisula* sp. Given *Asterias'* ability to exploit a range of other food resources, and the functional importance of bivalves in native systems (see Dame 1996), we predict wider population, community and ecosystem level effects. Overall, we suggest that the consequences of *A. amurensis* in the Derwent Estuary and the anticipated effects in areas outside the estuary warrant management efforts to control its spread and impact.

Acknowledgements

We thank the many volunteers and co-workers who braved the cold waters of southeast Tasmania, without whose help this project would not have been possible. Special thanks goes to Liz Turner, Graham Edgar, and Catriona Macleod for help with the identification of invertebrates. We thank Ron Thresher, Nic Bax, and three anonymous referees for constructive suggestions on the manuscript. This work was supported by funds from the CSIRO Marine Research Centre for Research on Introduced Marine Pests (awarded to CRJ) and the School of Zoology, University of Tasmania. This work was undertaken as part the senior author's Doctor of Philosophy degree at the University of Tasmania, which was supported by an Australian Postgraduate Award.

Note

1. http://www.fish.govt.nz/sustainability/ballast/ballast_strategy3.html#appendix1

References

Allmon RA and Sebens KP (1988) Feeding biology and ecological impact of an introduced nudibranch, *Tritonia plebia*, in New England. Marine Biology 99: 375–385

Barry JP and Dayton PK (1991) Physical heterogeneity and the organization of marine communities. In: Kolasa J and Pickett STA (eds) Ecological Heterogeneity, pp 270–320. Springer-Verlag, New York

Bax N (1999) Eradicating a dreissenid from Australia. *Dreissena!* 10: 1–4

Bennett B (1999) Healing the Derwent's murky blues. Ecos 100: 10–18

Bruce BD, Sutton CA and Lyne V (1995) Laboratory and field studies of the larval distribution and duration of the introduced seastar *Asterias amurensis* with updated and improved prediction of the species spread based on a larval dispersal model. Final report to the Fisheries Research and Development Corporation. CSIRO Division of Marine Research, Hobart, Australia

Buchanan JB (1966) The biology of *Echinocardium cordatum* (Echinodermata: Spatangoidea) from different habitats. Journal of the Marine Biological Association of the United Kingdom 46: 97–114

Buttermore RE, Turner E and Morrice MG (1994) The introduced northern Pacific seastar *Asterias amurensis* in Tasmania. Memoirs of the Queensland Museum 36: 21–25

Carlton JT and Geller JB (1993) Ecological roulette: the global transport of nonindigenous marine organisms. Science 261: 78–82

Clarke KR and Ainsworth M (1993) A method of linking multivariate community structure to environmental variables. Marine Ecology Progress Series 1993: 205–219

Cohen AN and Carlton JT (1996) Nonindigenous aquatic species in a United States estuary: a case history of the ecological and economic effects of biological invasions in the San Francisco and Delta region. Report to the United States Fisheries and Wildlife Service

Cohen AN and Carlton JT (1998) Accelerating invasion rate in a highly invaded estuary. Science 279: 555–558

Coles SL, DeFelice RC, Elderedge LG and Carlton JT (1999) Historical and recent introductions of non-indigenous marine species into Pearl Harbor, Oahu, Hawaiian Islands. Marine Biology 135: 147–158

Coughanowr C (1997) State of the Derwent Estuary: a review of environmental quality data to 1997. Supervising Scientists Report 129. Supervising scientist, Canberra

Culver C and Kuris A (2000) The apparent eradication of a locally introduced marine pest. Biological Invasions 2: 245–253

Dame RF (1996) Ecology of Marine Bivalves: An Ecosystem Approach. CRC Press, Boca Raton, Florida

Diamond J (1986) Overview: laboratory experiments, filed experiments, and natural experiments. In: Diamond J and Case T (eds) Community Ecology, pp 3–22. Harper and Row, New York

Elton CS (1958) The Ecology of Invasions by Animals and Plants. Methuen, London

Everett RA and Ruiz GM (1993) Coarse woody debris as a refuge from predation in aquatic communities, an experimental test. Oecologia 93: 475–486

Fernandes TF, Huxham M and Piper SR (1999) Predator caging experiments: a test of the importance of scale. Journal of Experimental Marine Biology and Ecology 241: 137–154

Fukuyama AK (1994) A review of the distribution and life history of *Asterias amurensis* on the northeast Pacific coast. Unpublished report to CSIRO and National Seastar Task Force, Hobart, Australia

Fukuyama AK and Oliver JS (1985) Sea star and walrus predation on bivalves in Morton Sound, Bering Sea, Alaska. Ophelia 24: 17–36

Furlani DM (1996) A guide to the introduced marine species in Australian waters. CSIRO Division of Fisheries, Hobart, Australia

Glasby T and Underwood A (1998) Determining the positions for control locations in environmental studies of estuarine marinas. Marine Ecology Progress Series 171: 1–14

Glasby TM (1997) Analysing data from post-impact studies using asymmetrical analysis of variance: a case study of epibiota on marinas. Australian Journal of Ecology 22: 448–459

Grannum RK, Murfet NB, Ritz DA and Turner E (1996) The distribution and impact of the exotic seastar, *Asterias amurensis* (Lutken), in Tasmania. In: The Introduced Northern Pacific Seastar, *Asterias amurensis*, in Tasmania, pp 53–138. Australian Nature Conservation Agency, Canberra

Griffiths CL, Hockey PAR, Schurink CV and Roux PJL (1992) Marine invasive aliens on South Africans shores: implication for community structure and trophic functioning. South African Journal of Marine Science 12: 713–722

Grosholz ED and Ruiz GM (1995) Spread and potential impact of the recently introduced European Green crab, *Carcinus maenas*, in central California. Marine Biology 122: 239–247

Grosholz E and Ruiz G (1996) Predicting the impact of introduced marine species: lessons from the multiple invasions of the European green crab *Carcinus maenas*. Biological Conservation 78: 59–66

Grosholz ED, Ruiz GM, Dean CA, Shirley KA, Maron JL and Connors PG (2000) The impacts of a nonindigenous marine predator in a California Bay. Ecology 81: 1206

Hatanaka M and Kosaka M (1959) Biological studies on the population of the starfish, *Asterias amurensis*, in Sendai Bay. Tohoku Journal of Agricultural Research 9: 159–178

Hewitt CL, Campbell ML, Thresher RE and Martin RB (1999) Marine biological invasions of Port Phillip Bay, Victoria. CSIRO Marine Research, Hobart, Australia

Hughes RN, Peer DL and Mann KH (1972) Use of multivariate analysis to identify functional components of the benthos in St Margaret's Bay, Novia Scotia. Limnology and Oceanography 17: 111–121

Johnson D (1994) Seastar fight gains momentum. Research 53: 25–29

Jones M (1991) Marine organisms transported in ballast water: a review of the Australian scientific position. Department of Primary Industries and Energy, Bureau of Rural Resources, AGPS, Canberra

Kareiva P (1994) Higher order interactions as a foil to reductionist ecology. Ecology 75: 1527–1528

Keough MJ and Mapstone BD (1995) Protocols for designing marine ecological monitoring programs associated with BEK mills. National Pulp Mills Research Program Technical Report No 11. CSIRO, Canberra

Keough MJ and Mapstone BD (1997) Designing environmental monitoring for pulp mills in Australia. Water Science and Technology 35: 397–404

Kim YS (1969) Selective feeding on the several bivalve molluscs by starfish, *Asterias amurensis*. Bulletin of the Facility of Fisheries Hokkaido University 19: 244–249

Lipcius RN and Hines AH (1986) Variable functional responses of a marine predator in dissimilar homogenous habitats. Ecology 67: 1361–1371

Lockhart SJ (1995) Feeding biology of the introduced sea star, *Asterias amurensis* (Lutken) in Tasmania (Echinodermata Asteroidea). Department of Zoology. Honours thesis, University of Tasmania, Hobart, Australia

Lockhart SJ and Ritz DA (2001a) Size selectivity and energy maximisation of the introduced seastar, *Asterias amurensis* (Lutken), in Tasmania, Australia. Papers and Proceedings of the Royal Society of Tasmania 135: 35–40

Lockhart SJ and Ritz DA (2001b) Preliminary observations of the feeding periodicity and selectivity of the introduced seastar, *Asterias amurensis* (Lutken), in Tasmania, Australia. Papers and Proceedings of the Royal Society of Tasmania 135: 25–33

Lodge DM (1993) Biological Invasions: lessons for ecology. Trends in Ecology and Evolution 8: 133–137

Lodge DM, Stein RA and Brown KM (1998) Predicting impact of freshwater exotic species on native biodiversity: challenges in spatial scaling. Australian Journal of Ecology 23: 53–67

McLoughlin R and Bax N (1993) Scientific discussions in Japan and Russia on the northern Pacific seastar. Unpublished report. CSIRO Division of Marine Research, Hobart, Australia

McLoughlin R and Thresher R (1994) The north Pacific seastar: Australia's most damaging pest? Search 25: 69–71

Menge BA (1982) Effects of feeding on the environment: Asteroidea. In: Janguoux M and Lawrence JM (eds) Echinoderm Nutrition, pp 521–554. A.A. Balkema, Rotterdam, The Netherlands

Morrice MG (1995) The distribution and ecology of the introduced northern Pacific seastar, *Asterias amurensis* (Lutken), in Tasmania. In: The Introduced Northern Pacific Seastar, *Asterias amurensis*, in Tasmania, pp 1–47. Australian Nature Conservation Agency, Canberra

Murphy N and Evans B (1998) Genetic origin of Australian populations of *Asterias amurensis*. Proceedings of a meeting on the biology and management of the introduced seastar, *Asterias amurensis*, in Australian waters pp 22–25. CSIRO Division of Marine Research, Hobart, Australia

Nichols FH, Thompson JK and Schemel LE (1990) Remarkable invasion of San Francisco Bay (California, USA) by the Asian clam *Potamocorbula amurensis*. II. Displacement of a former community. Marine Ecology Progress Series 66: 95–101

Nojima S, Soliman FA, Kondo Y, Kuwano Y, Nasu K and Kitajima C (1986) Some notes of the outbreak of the sea star *Asterias amurensis versiclor* Sladen, in the Ariake Sea, western Kyshu. Publication of the Amakusa Marine Biology Laboratory Kyushi University 8: 89–112

Osenberg CW and Schmitt RJ (1996) Detecting ecological impacts caused by human activities. In: Schmitt RJ and Osenberg CW (eds) Detecting Ecological Impacts: Concepts and Applications in Coastal Habitats, pp 3–16. Academic Press, San Diego, California

Parker IM, Simberloff D, Lonsdale D, Goodel K, Wonham M, Kareiva PM, Williamson MH, Von Holle B, Moyle PB, Byers JE and Goldwasser L (1999) Impact: toward a framework for understanding the ecological effects of invaders. Biological Invasions 1: 3–19

Pimentel D, Lach L, Zuniga R and Morrison D (2000) Environmental and economic costs of nonindigenous species in the United States. BioScience 50: 53–64

Pollard DA and Hutchings PA (1990a) A review of exotic marine organisms introduced to the Australian region. I. Fishes. Asian Fisheries Science 3: 205–221

Pollard DA and Hutchings PA (1990b) A review of exotic marine organisms introduced to the Australian region. 2. Invertebrates and Algae. Asian Fisheries Science 3: 222–250

Ross DJ (2001) Impact of the northern Pacific seastar *Asterias amurensis* on soft sediment assemblages, including commercial species, in southeast Tasmania. PhD dissertation, University of Tasmania, Hobart, Australia

Ross DJ, Johnson CR and Hewitt CL (2002) Impact of the introduced seastar *Asterias amurensis* on survivorship of juvenile commercial bivalves *Fulvia tenuicostata*. Marine Ecology Progress Series 241: 99–112

Ross DJ, Johnson CR and Hewitt CL (2003) Variability in the impact of an introduced predator (*Asterias amurensis*: Asteroidea) on soft sediment assemblages. Journal of Experimental Marine Biology and Ecology 288: 257–278

Ross DJ, Johnson CR and Hewitt CL (in press) Interaction and impacts of two introduced species on a soft sediment marine assemblage in SE Tasmania. Marine Biology

Rhoads DC (1974) Organism-sediment relations on the muddy seafloor. Oceanography and Marine Biology: an Annual Review 12: 263–300

Ruiz GM, Fofonoff P, Hines AH and Grosholz ED (1999) Non-indigenous species as stressors in estuarine and marine communities: assessing invasion impacts and interactions. Limnology and Oceanography 44: 950–972

Sanders HL (1958) Benthic studies in Buzzard's Bay. I. Animal–sediment relationships. Limnology and Oceanography 3: 245–258

Schmitt RJ and Osenberg CW (eds) (1996) Detecting Ecological Impacts: Concepts and Applications in Coastal Habitats. Academic Press, San Diego, California

Shushkina E and Musayeva E (1990) Structure of the plankton community from the Black Sea and its changes as a result of the introduction of a ctenophore species. Oceanology 30: 225–228

Simberloff D and Holle BV (1999) Positive interaction of non-indigenous species: invasional meltdown? Biological Invasions 1: 21–32

Skilleter GA (1994) Refuges from predation and the persistence of estuarine clam populations. Marine Ecology Progress Series 109: 29–42

Sloan NA (1980) Aspects of the feeding biology of asteroids. Oceanography and Marine Biology: an Annual Review 18: 57–124

Thresher RE (2000) Key threats from marine bioinvasions: a review of current and future issues. In: Pederson J (ed) Proceedings of the First National Conference of Marine Bioinvasions. MIT Press, Boston

Thrush SF (1999) Complex role of predators in structuring soft-sediment macrobenthic communities: implications of changes in spatial scale for experimental studies. Australian Journal of Ecology 24: 344–354

Thrush SF, Cummings VJ, Dayton PK, Ford R, Grant J, Hewitt JE, Hines AH, Lawrie SM, Pridmore RD, Legendre P, Mcardle BH, Schneider DC, Turner SJ, Whitlatch RB and Wilkinson MR (1997) Matching the outcome of small-scale density manipulation experiments with larger scale patterns an example of bivalve adult/juvenile interactions. Journal of Experimental Marine Biology and Ecology 216: 153–169

Thrush SF, Hewitt JE, Cummings VJ, Green MO, Funnell GA and Wilkinson MR (2000) The generality of field experiments: interactions between local and broad-scale processes. Ecology 81: 399–415

Turner E (1992) A northern Pacific seastar, *Asterias amurensis*, in Tasmania. Australian Marine Science Bulletin 120: 18–19

Underwood AJ (1991) Beyond BACI: experimental designs for detecting human environmental impacts on temporal variations in natural populations. Australian Journal of Marine and Freshwater Research 42: 569–587

Underwood AJ (1992) Beyond BACI: the detection of environmental impacts on populations in the real, but variable, world. Journal of Experimental Marine Biology and Ecology 161: 145–178

Underwood AJ (1996) Detection, interpretation, prediction and management of environmental disturbances: some roles for experimental marine ecology. Journal of Experimental Marine Biology and Ecology 200: 1–27

Virnstein RW (1977) The importance of predation by crabs and fishes on benthic infauna in Chesapeake Bay. Ecology 58: 119–127

Vitousek PM, D'Antonio CM, Loope LL and Westbrooks R (1996) Biological invasions as global environmental change. American Scientist 84: 468–478

Willan RC, Russell BC, Murfet NB, Moore KL, McEnnulty FR, Horner SK, Hewitt CL, Dally GM, Campbell ML and Bourke ST (2000) Outbreak of *Mytilopsis sallei* (Récluz, 1849) (Bivalvia; Dressenida) in Australia. Molluscan Research 20(2): 25–30

Williamson M (1996) Biological Invasions. Chapman & Hall, London

Woodin SA (1974) Polychaete abundance patterns in a marine soft sediment environment: the importance of biological interactions. Ecological Monographs 44: 171–87

Woodin SA (1978) Refuges, disturbance and community structure: a marine soft-bottom example. Ecology 59: 274–284

Biological Invasions **5:** 23–31, 2003.

Native species vulnerability to introduced predators: testing an inducible defense and a refuge from predation

W. Lindsay Whitlow[1,2,*], Neil A. Rice[2] & Christine Sweeney[2]
[1] *Department of Biology/Environmental Studies, Bowdoin College, Brunswick, ME 04011, USA;*
[2] *Wells National Estuarine Research Reserve, 342 Laudholm Farm Rd., Wells, ME 04090, USA;*
Author for correspondence (e-mail: lwhitlow@bowdoin.edu; fax: +1-207-725-3405)

Received 11 July 2001; accepted in revised form 6 June 2002

Key words: behavior, depth, *Carcinus maenas*, inducible defense, introduced predator, invasive species, *Mya arenaria*, native species, predator avoidance, refuge

Abstract

To manage the impacts of biological invasions, it is important to determine the mechanisms responsible for the effects invasive species have on native populations. When predation by an invader is the mechanism causing declines in a native population, protecting the native species will involve elucidating the factors that affect native vulnerability. To examine those factors, this study measured how a native species responded to an introduced predator, and whether the native response could result in a refuge from predation. Predation by the green crab, *Carcinus maenas*, has contributed to the decline in numbers of native soft-shell clams, *Mya arenaria*, and efforts to eradicate crabs have proven futile. We tested how crab foraging affected clam burrowing, and how depth in the sediment affected clam survival. Clams responded to crab foraging by burrowing deeper in the sediment. Clams at shallow depths were more vulnerable to predation by crabs. Results suggest soft-shell clam burrowing is an inducible defense in response to green crab predation because burrowing deeper results in a potential refuge from predation by crabs. For restoring the native clam populations, tents could exclude crabs and protect clams, but when tents must be removed, exposing the clams to cues from foraging crabs should induce the clams to burrow deeper and decrease vulnerability. In general, by exposing potential native prey to cues from introduced predators, we can test how the natives respond, identify whether the response results in a potential refuge, and evaluate the risks to native species survival in invaded communities.

Introduction

With increasing rates of species introductions around the world, there is a need to understand the effects that invasive species can have on native species (Carlton 1996; Carlton and Geller 1993; Lodge 1993). Many studies are currently documenting patterns of invasive species establishment and identifying traits common among successful invaders (Kolar and Lodge 2001; Rejmanek 2000; Byers 1999; Cohen et al. 1995). However, only recently have studies begun to test the mechanisms responsible for the successful invasion of species into new habitats, and few studies have addressed the mechanisms underlying declines in native species after invasion (Christian 2001; McDonald et al. 2001; Byers 2000; Grosholz et al. 2000; Ruiz et al. 1997). This gap needs to be filled if we are to develop successful strategies for protecting or restoring native populations.

Native species can be vulnerable to introduced predators because the predators employ foraging methods that the natives have not encountered or the predators are introduced where no predators existed. Classic examples are the dramatic decline in native birds on Guam after the introduction of the brown tree snake, *Boiga irregularis* (Fritts and Rodda 1998), and the loss of native cichlids in Lake Victoria after the introduction of Nile Perch, *Lates niloticus* (Kitchell et al. 1997; Witte et al. 1992). In these cases, the native species were endemic to islands or lakes, but introduced predators can pose a serious threat to more geographically widespread native species because of an increase in overall predator abundance. After these introduced predators become established, they can be difficult to eradicate, so the most viable management option may be to focus on protecting the native species.

To protect a native species when predation is a mechanism responsible for population decline, efforts should focus on identifying factors that affect native vulnerability. Recent studies have documented the prevalence of changes in prey behavior and morphology as inducible defenses in response to predators (Trussell and Smith 2000; Leonard et al. 1999; Tollrian and Dodson 1999; Trussell 1996). For example, to decrease vulnerability to predators, the larvae of some anuran species avoid open water or increase tail size relative to body size (Peacor and Werner 2001; Relyea 2001). Also, *Daphnia pulex* migrate deeper in the water column or grow protective spines in response to predators (Dodson et al. 1997; Tollrian 1995). These inducible defenses could be a significant factor in the survival of native species in invaded communities since natives capable of detecting and responding to introduced predators could be less vulnerable.

Optimal foraging theory predicts that predation will depend upon the costs and benefits involved in acquiring and consuming prey (Stephens and Krebs 1986). As costs to the predator increase through higher energy expenditure or threat of predation, individual prey may become less vulnerable (Meire and Ervynck 1986). Therefore, a refuge from predation may exist under conditions when costs exceed benefits for predators searching for and capturing prey. Many species attain a refuge from predation through growth since their larger size increases the energy a predator must expend to capture the prey (Nilsson 2001; Tollrian and Dodson 1999; Olson 1996; Anholt and Werner 1995). Other species attain a refuge through morphological (Bronmark and Miner 1992), chemical (Stachowicz and Hay 1999), or behavioral defenses (Banks 2001; Peacor and Werner 2001). To protect a native species in an invaded community, it would be useful to determine how the native species responds to the introduced predator, and whether that response could result in a potential refuge from predation.

Invasion of the green crabs

In the Northwest Atlantic, predation by the introduced green crab, *Carcinus maenas*, has contributed to dramatic declines in native soft-shell clams, *Mya arenaria* (Glude 1955). An increase in green crab abundance in the 1940s and 1970s in the region, coupled with over-harvesting and poor water quality (Dow 1957) led to an even more dramatic decline in the clam population in Maine (Beal 1994). Studies have correlated increases in green crab abundance with declines in soft-shell clam populations (Welch 1968). Due to the value of the soft-shell clam fishery to local communities, projects to restore clams are currently underway (Beal et al. 1995). Initially much effort was expended on eradicating crabs, but the crabs' planktonic larvae allow populations to become re-established too rapidly for current efforts to be successful. Therefore, a goal of this study was to provide an effective strategy for protecting clams under the assumption that crab populations will persist in these communities. To address that goal, preliminary experiments were conducted to determine whether the clams respond to the crabs, and whether the response could provide a refuge from predation.

Inducible defense

Soft-shell clams burrow in the sediment by ejecting water through the pedal gape and using their foot as an anchor (Checa and Cadee 1997). Unlike other clam species, the foot of soft-shell clams is small relative to its body size. Therefore, the clams burrow predominantly vertically, meaning their depth may change, but their spatial location on the mudflat does not change dramatically (Van Dover and Kirby-Smith 1979). In laboratory experiments, soft-shell clams increased depth in the sediment in response to treatments simulating red rock crab, *Cancer productus*, activity (Zaklan and Ydenberg 1997). This study tested whether soft-shell clams in the field burrow deeper in response to green crabs as a potential inducible defense.

Refuge from predation

For crabs, foraging costs include time spent searching and excavating prey, so deeper clams should be less vulnerable (Smith et al. 1999; Behrens-Yamada and Boulding 1996; Ebersole and Kennedy 1995; Skilleter 1994; Moody and Steneck 1993; Elner 1978; Virnstein 1977). For low densities of clams (*Mya arenaria* and *Macoma balthica*), depth was an important refuge from predation by blue crabs (*Callinectes sapidus*) in

Chesapeake Bay (Eggleston et al. 1992). Soft-shell clams deeper in the sediment were also less vulnerable to *Cancer productus* predation on the Pacific coast (Zaklan and Ydenberg 1997). This study tested whether depth in the sediment could be a refuge for soft-shell clams from green crab predation.

We tested two specific hypotheses on how green crabs could affect soft-shell clams: (1) If clams respond to crabs, then clams exposed to crab foraging will burrow deeper in the sediment; and (2) If greater depth in the sediment is a refuge from predation by crabs, then clams at shallow depths will be more vulnerable than deeper clams.

Methods

Species descriptions

Soft-shell clams, *Mya arenaria*, are native to the Atlantic coast of North America, and their range extends from the Arctic Ocean to Cape Hatteras, North Carolina (Van Dover and Kirby-Smith 1979; Dow 1957). Adults live buried in intertidal mudflats with their siphon extended into the water column allowing the intake of food and oxygen, and the removal of waste (Van Dover and Kirby-Smith 1979; Dow 1957). When clams are removed from the sediment, their thin shell and open valves from weak adductor muscles make them highly vulnerable to digging predators (Boulding 1984).

Green crabs, *Carcinus maenas*, arrived on the Atlantic coast of North America more than a century ago from Europe, and their range now extends from Maryland to Prince Edward Island (Levinton 1982). Crabs were also introduced into South Africa, Australia, Tasmania, Japan, and most recently the Pacific coast of North America (Ruiz et al. 1997; Grosholz and Ruiz 1996). Ballast water is the likely vector for the current global distribution of green crabs (Carlton 1996; Cohen et al. 1995). The ability of green crabs to tolerate wide salinity (5–30 ppt) and temperature (5–30 °C) ranges enables populations to become established in most temperate intertidal areas (Grosholz and Ruiz 1995; Levinton 1982). Green crabs are omnivorous predators, but prefer mollusks such as snails and clams (Hughes and O'Brien 2001; Maaski and Guillou 1999; van der Veer et al. 1998; Rangeley and Thomas 1987; Elner 1981; Hughes and Elner 1979).

Field site

All field experiments and organism collections were conducted on mudflats within the Little River estuary at the Wells National Estuarine Research Reserve (WNERR) in Wells, Maine, USA. Field experiments and collections took place from May through September in 1998 and 1999.

Clam and crab collection

Soft-shell clams were collected within 0.25 m^2 quadrats randomly selected within discrete mudflat areas. Clam density, size, and depth in the sediment were measured. Green crabs were collected using unbaited pitfall traps constructed of five-gallon plastic buckets filled with water and buried in the sediment so the opening was flush with the mud surface. Since green crabs are poor swimmers, they remained trapped within the buckets after falling in. Crab collection periods took place twice a month for four consecutive days. During collection periods, traps were checked every 24 h and all crabs were removed. We measured crab abundance (number collected/day), size (carapace width), abdomen color (green vs. red), sex, and larger claw (width and length).

Inducible defense

To test how clams respond to crab foraging, a field experiment manipulating clam exposure to crab foraging was conducted. Inside 0.25 m^2 plots on the mudflat, five clams (55–65 mm shell length) were buried at a depth of 10 cm from the mud surface to the siphonal end of the shell. Clams were buried 10 cm deep in the sediment to reflect the mean depth measured for clams in that size range during population surveys. Plots were covered with a tent of rigid 3 mm Vexar plastic mesh (50 cm wide × 50 cm long × 20 cm tall) with the bottom edge buried 10 cm into the mud on all sides and the top of the tent standing 10 cm above the mud surface. Clams were left for one week to acclimate to their depths in the sediment.

In crab exclosure treatments, mesh tents were left intact over the plots. In treatments open to crab foraging, mesh tents were cut so that a 3 cm tall fence of mesh remained around the plot. The mesh fences were low enough for crabs to climb over easily and attempted to control for effects of mesh on water flow to the clams. In crab enclosure treatments, crabs

(45–55 mm carapace width) were placed within the mesh tents over the plots. After 90 days, mesh tents and crabs were removed from the plots. Clam depths were measured by placing a meter stick horizontally across the mud surface and measuring the vertical distance from the meter stick to the siphonal end of each clam shell. Since *Mya arenaria* predominantly burrows vertically in the sediment, vertical distance from the sediment surface could be used as an accurate measure of clam depth. There were five replicates of each experimental treatment, with treatments randomly distributed across a grid of plots. Plots were separated by 2 m within the 50 m^2 grid. Data was analyzed using a one-way ANOVA with crab treatment as the fixed factor and clam depth as the dependent factor.

Refuge from predation

To test how clam depth affected survival, a field experiment manipulating clam depth and exposure to crab predation was conducted. Four clams (55–65 mm shell length) were placed into 20 cm diameter plastic flowerpots. The pots were filled with sediment and placed into the mudflats flush with the mud surface. In crab exclosure treatments, pots were covered with a tent of rigid 3 mm Vexar plastic mesh. Tents were assembled as cylinders (20 cm diameter × 15 cm tall) with tops. In treatments open to crab predation, 3 cm tall cylinders without tops were placed around the pots. For all pots, the bottom edges of the mesh tents attached around the pots with rubber bands. In shallow depth treatments, maximum clam depth was restricted to 5 cm by shortening pot height. In deep treatments, the bottoms of pots were removed to allow clams to burrow without restriction. There were six replicates of each experimental treatment, with treatments randomly distributed across a grid of pots on the mudflat. Pots were separated by 1 m within the 24 m^2 grid. After 30 days, the number of surviving clams in each pot was measured. Broken clam shells or missing clams were considered victims of predation. Data were analyzed using a two-way ANOVA with crab treatment and clam depth as fixed factors, and the proportion of clams surviving as the dependent factor. Proportional data were arcsine transformed to control for normality.

For both experiments, Tukey's *post-hoc* analysis was used to test for differences among the treatments using SYSTAT 9.

Results

Clam and crab size distributions

The most abundant size class among collected clams was between 50 and 70 mm shell length (Figure 1). The most abundant size class among collected crabs was between 40 and 60 mm carapace width (Figure 1). Both crab and clam size distributions were slightly bimodal, with smaller peaks for clam shell lengths at 21–30 mm and crab carapaces at 0–20 mm. There was a positive correlation between clam shell length and depth (Figure 2, Pearson's correlation = 0.697, $P < 0.001$).

Inducible defense

Clams responded to crab foraging by burrowing deeper in the sediment (Figure 3). There was a significant effect of crab treatment on clam depth ($P < 0.05$, ANOVA). Clams buried significantly deeper in crab enclosures compared to exclosure treatments (Table 1, $P < 0.05$, Tukey's *post-hoc*). There were no significant differences in clam depths between open and exclosure ($P = 0.762$) or enclosure treatments ($P = 0.191$). The data met assumptions of normality and equal variance (Residuals: Skewness = −0.179, Kurtosis = 0.097, Lilliefor's test $P = 0.713$, Levene's test $P = 0.426$). Clam survival did not significantly differ between crab treatments (Table 1, $P = 0.220$),

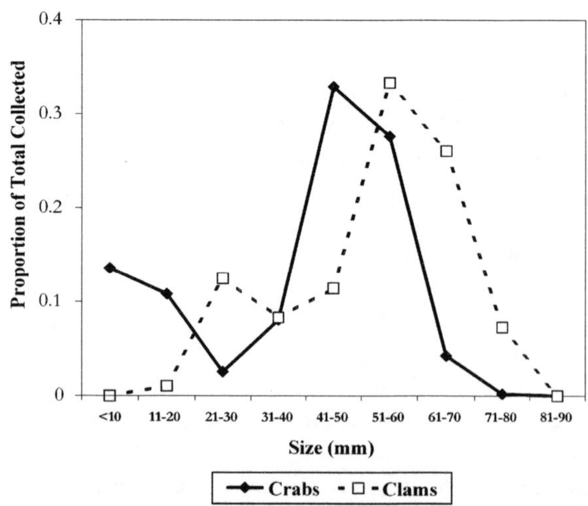

Figure 1. Size distribution of green crabs ($N = 471$) and soft-shell clams ($N = 192$) collected May–August 1998 in the Little River estuary, Wells, Maine. Carapace widths of green crabs and shell length of soft-shell clams were measured (mm).

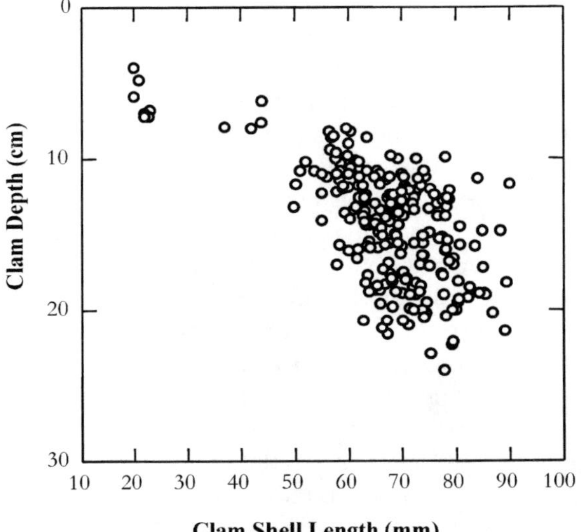

Figure 2. The positive correlation between clam shell length and depth in the sediment (Pearson's correlation = 0.697, $P < 0.001$).

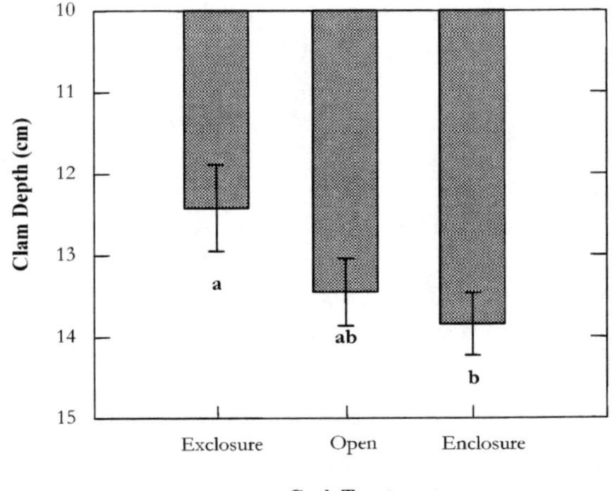

Figure 3. Clams responded to crab foraging by burrowing deeper in the sediment ($P < 0.05$, ANOVA). Error bars represent ±1 S.E.

and the data met assumptions of normality and equal variance (Residuals: Skewness = −0.382, Kurtosis = −0.909, Lilliefor's test $P = 0.114$, Levene's test $P = 1.000$).

Refuge from predation

Clams at shallow depths were more vulnerable to predation (Figure 4). There was a significant interaction between crab and clam depth treatments ($P < 0.001$, ANOVA). Clam survival was high in crab exclosures

Table 1. Effects of crab foraging treatments on clam burrowing depth and survival.

	Exclosure	Open	Enclosure	P
Depth	12.419^{a}	13.451^{ab}	13.843^{b}	>0.05
(mm)	(±0.467)	(±0.359)	(±0.334)	
Proportion of	0.900	0.760	0.880	0.220
clams surviving	(±0.055)	(±0.060)	(±0.058)	

Superscript letters in common indicate no significant difference between treatments.

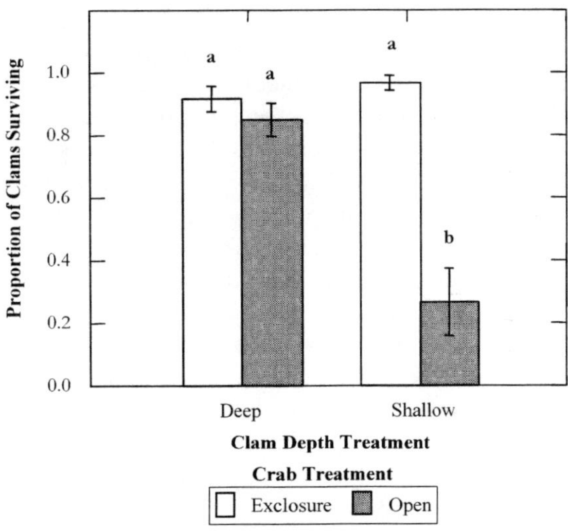

Figure 4. When unprotected, clams at shallow depths were more vulnerable to predation after 30 days ($P < 0.001$, ANOVA). Error bars represent ±1 S.E.

whether clams were buried deep (0.917 ± 0.039) or shallow (0.967 ± 0.022) in the sediment ($P = 0.868$, Tukey's *post-hoc*). In plots open to crabs, survival was low for shallow clams (0.267 ± 0.102), but survival was high for deep clams (0.850 ± 0.050) ($P < 0.001$, Tukey's *post-hoc*). Survival did not significantly differ between clams in crab exclosures and deep clams in plots open to crabs ($P = 0.938$ and $P = 0.539$, Tukey's *post-hoc*). The arcsin transformed data met assumptions of normality and equal variance (Residuals: Skewness = 0.478, Kurtosis = 0.948, Lilliefor's test $P = 0.160$, Levene's test $P = 1.000$).

Discussion

Inducible defense

Deeper burrowing by clams in response to foraging crabs indicates the native clams are capable of changing behavior to avoid the introduced predator (Figure 3). This change in soft-shell clam behavior suggests deeper burrowing may be an inducible defense against green crabs. Laboratory studies have demonstrated clam depth changes in response to predation cues (J. Juhasz, unpub. Master's thesis, 2000; Smith et al. 1999; Zaklan and Ydenberg 1997). The transition in clam depth between exclosure, open, and enclosure treatments suggests that clam burrowing in the field can differ between areas that differ in exposure to crab foraging.

Crab exposure altered clam behavior, but we did not identify the specific cue that induced the response. Clams in exclosures may have not burrowed as deep because those plots eliminated the tactile cues of crab foraging. Alternatively, clams could have been more shallow in exclosure plots because the intensity of chemical cues was lower. Chemical cues that could be released from feeding crabs include digestive enzymes, hemolymph, and alarm cues (Hazlett 1996). Clams have been found to respond to chemical cues released during crab predation when isolated from any tactile cues (Whitlow 2002).

Refuge from predation

For soft-shell clams, it was depth in the sediment that provided a refuge from green crab predation, not their larger size (Figure 4). This differs from other species where larger body size can provide a refuge from predation when predators cannot capture the larger prey (Nilsson 2001; Tollrian and Dodson 1999; Anholt and Werner 1995). Large green crabs (45–55 mm carapace width) were observed breaking the shells of large soft-shell clams and consuming them (Whitlow, pers. obs.). Hence, large soft-shell clams, despite their size, were vulnerable to green crab predation if their depth was shallow enough where crabs could excavate them.

The clam size distribution skewed to larger sizes (Figure 1) may be due to predation by crabs on smaller clams that cannot burrow as deep in the sediment because their siphons are shorter (van der Veer et al. 1998). Alternatively, the presence of lower peaks at small sizes of both clams and crabs suggests peaks in the size distributions may represent recruitment cohorts. However, low survival of juvenile clams has been found to be a bottleneck in clam population persistence on invaded mudflats (Maaski and Guillou 1999; van der Veer et al. 1998; Wilson 1991).

Prey species that change behavior in response to a predator can face a trade-off between predator avoidance and feeding (Peacor and Werner 2001; Anholt and Werner 1995). For soft-shell clams, deeper burrowing may increase predator avoidance but decrease feeding efficiency (de Goeij and Luttikhuizen 1998; Zaklan and Ydenberg 1997). The positive correlation between clam shell length and depth (Figure 2) suggests clams may burrow deeper to avoid predators, but siphon length may restrict clam depth. For deeper clams to continue feeding, their siphons must be long enough to reach the surface. In addition, feeding efficiency declines at greater depths because more energy must be expended by clam muscles to inhale water through the siphon to the gills (de Goeij and Luttikhuizen 1998; Zaklan and Ydenberg 1997). Studies suggest tissue and shell growth can be uncoupled in soft-shell clams. Therefore, deeper clams could grow longer siphons, but at the expense of shell growth (Lewis and Cerrato 1997). Deeper burrowing may provide a short-term benefit of increased survival, but there may be long-term costs in decreased body growth, fecundity, or longevity because of lower feeding efficiency (de Goeij and Luttikhuizen 1998). These long-term costs may explain the decline in soft-shell clam populations despite the clam's ability to burrow deeper to avoid the green crabs. Future investigations into the impact of introduced predators on native prey should measure parameters beyond short-term survival because of potential long-term effects associated with the costs of native responses.

Implications for clam restoration

Because burrowing depth is important to clam survival, field methods for protecting native clams should incorporate both exclusion techniques (e.g. mesh tents) and cues to induce effective predator avoidance (e.g. tactile or chemical cues). In clam restoration projects, mesh tents are being used more frequently to protect soft-shell clams (Beal 1994). Results from this study show mesh tents effectively protect clams, but excluding crabs with tents will result in clams that do not burrow as deep. If tents are removed after one summer, which is common, those shallow clams could be more vulnerable to crab predation. To insure clams are protected, and induce clams to burrow deeper, chemical cues from crab predation could be released beneath the tents. Juvenile clams are predominantly used in these restoration projects, and juveniles have been found to burrow deeper and grow longer siphons in response to chemical cues from crab predation (Whitlow, 2002). The longer-term costs of inducing deeper burrowing have yet to be measured.

Implications for native species protection

A potential method for rapidly assessing which native species may be most dramatically affected by an invasive species could be to expose the natives to cues released by the invader and measure native responses. For example, if the invader could be a predator on multiple native species, it would be useful to measure how each native species responds to cues released during predation by the invader. Based on short experiments, natives that responded in ways that could reduce their vulnerability to the invader would be predicted to be less affected by the invasion in the short-term. Then, management could be initially directed towards protecting those natives that did not respond as effectively to the invader cues. Further studies would remain necessary to determine how the invader could affect the native species over the long-term.

Conclusions

Depth in the sediment was identified as a factor important for decreasing vulnerability of native soft-shell clams in estuaries with introduced green crabs. Clams responded to crabs by burrowing deeper in the sediment. Survival of clams deeper in the sediment was higher than clams at more shallow depths. This study is intended to provide preliminary information for the development of more effective strategies for protecting soft-shell clams. The results from this study suggest clam depth is an important parameter to measure when monitoring clam populations in invaded estuaries or during restoration projects. In addition, excluding crabs is initially effective for protecting clams, but we may promote clam survival after tent removal by inducing clams to burrow deeper while protected. Also, soft-shell clam traits beyond survival should be monitored to measure long-term impacts of the green crabs on clam growth and reproduction.

In general, by exposing potential native prey to cues from introduced predators, we can test how the natives respond, identify whether the response results in a potential refuge, and evaluate the risks to native species survival in invaded communities.

Acknowledgements

This research was conducted under an award from the Estuarine Reserves Division, Office of Ocean and Coastal Resource Management, National Ocean Service, National Oceanographic and Atmospheric Administration. The Helen Olsen Brower Memorial Fellowship from the University of Michigan also provided funding support. Block grant funding from the University of Michigan Department of Biology assisted with initial research costs. Thanks to the entire staff of Wells National Estuarine Research Reserve for assistance, particularly M. Dionne, S. Orringer, R. MacKenzie, T. Smith, and N. Bayse. Thanks to B. Hazlett, R. MacKenzie, C. Mullen, P. Ewanchuk, S. MacCauley, K. Judd, J. Pederson, and four anonymous reviewers for reviewing the manuscript. Thanks to B. Beal, B. Walton, M. Dionne, S. Orringer, B. Hazlett, and S. Schueller, for help with developing ideas.

References

Anholt BR and Werner EE (1995) Interaction between food availability and predation mortality mediated by adaptive behavior. Ecology 76: 2230–2234

Banks PB (2001) Predation-sensitive grouping and habitat use by eastern grey kangaroos: a field experiment. Animal Behavior 61: 1013–1021

Beal BF (1994) Biotic and abiotic factors influencing growth and survival in wild and cultured individuals of the soft-shell clam, *Mya arenaria* L., in Eastern Maine. PhD thesis, University of Maine, Orono, Maine

Beal BF, Lithgow CD, Shaw DP, Renshaw S and Ouellette D (1995) Overwintering hatchery-reared individuals of the soft-shell clam, *Mya arenaria* L. – a field-test of site, clam size, and intraspecific density. Aquaculture 130: 145–158

Behrens-Yamada S and Boulding EG (1996) The role of highly mobile crab predators in the intertidal zonation of their gastropod prey. Journal of Experimental Marine Biology and Ecology 204: 59–83

Boulding EG (1984) Crab-resistant features of shells of burrowing bivalves: decreasing vulnerability by increasing handling time. Journal of Experimental Marine Biology and Ecology 76: 201–223

Bronmark C and Miner JG (1992) Predator-induced phenotypical change in body morpholgy in crucian carp. Science 258(5086): 1348–1350

Byers JE (1999) The distribution of an introduced mollusc and its role in the long-term demise of a native confamilial species. Biological Invasions 1: 339–352

Byers JE (2000) Competition between two estuarine snails: implications for invasions of exotic species. Ecology 81(5): 1225–1239

Carlton JT (1996) Pattern, process, and prediction in marine invasion ecology. Biological Conservation 78: 97–106

Carlton JT and Geller JB (1993) Ecological roulette: the global transport of nonindigenous marine organisms. Science 261: 78–82

Christian CE (2001) Consequences of a biological invasion reveal the importance of mutualism for plant communities. Nature 413: 635–639

Checa AG and Cadee GC (1997) Hydraulic burrowing in the bivalve *Mya arenaria* linnaeus (Myoidea) and associated ligamental adaptations. Journal of Molluscan Studies 63: 157–171

Cohen AN, Carlton JT and Fountain MC (1995) Introduction, dispersal and potential impacts of the green crab *Carcinus maenas* in San Francisco Bay, California. Marine Biology 122: 225–237

Dodson SI, Tollrian R and Lampert W (1997) Daphnia swimming behavior during vertical migration. Journal of Plankton Research 19(8): 969–978

Dow RL (1957) The Maine clam, *Mya arenaria*. A bulletin of the Department of Sea and Shore Fisheries, State House, Augusta, Maine

Ebersole EL and Kennedy VS (1995) Prey preferences of blue crabs *Callinectes sapidus* feeding on three bivalve species. Marine Ecology Progress Series 118: 167–177

Eggleston DB, Lipcius RN and Hines AH (1992) Density-dependent predation by blue crabs upon infaunal clam species with contrasting distribution and abundance patterns. Marine Ecology Progress Series 85: 55–68

Elner RW (1981) Diet of green crab *Carcinus maenas* (L.) from Port Herbert, Southwestern Nova Scotia. Journal of Shellfish Research 1: 89–94

Elner RW (1978) The mechanics of predation by the shore crab, *Carcinus maenas* (L.), on the edible mussel, *Mytilus edulis* L. Oecolgia 36: 333–344

Fritts TH and Rodda GH (1998) The role of introduced species in the degradation of island ecosystems: a case history of Guam. Annual Review of Ecology and Systematics 29: 113–140

Glude JB (1955) The effects of temperature and predators on the abundance of the soft-shell clam, *Mya arenaria*, in New England. Transactions of the American Fisheries Society 84: 13–26

de Goeij P and Luttikhuizen P (1998) Deep-burying reduces growth in intertidal bivalves: field and mesocosm experiments with *Macoma balthica*. Journal of Experimental Marine Biology and Ecology 228: 327–337

Grosholz ED, Ruiz GM, Dean CA, Shirley KA, Maron JL and Connors PG (2000) The impacts of a nonindigenous marine predator in a California bay. Ecology 81: 1206–1224

Grosholz ED and Ruiz GM (1996) Predicting the impact of introduced marine species: lessons from the multiple invasions of the European green crab *Carcinus maenas*. Biological Conservation 78: 59–66

Grosholz ED and Ruiz GM (1995) Spread and potential impact of the recently introduced European green crab, *Carcinus maenas*, in central California. Marine Biology 122: 239–247

Hazlett BA (1996) Organisation of hermit crab beaviour: responses to multiple chemical inputs. Behaviour 133: 619–642

Hughes RN and O'Brien N (2001) Shore crabs are able to transfer learned handling skills to novel prey. Animal Behaviour 61: 711–714

Hughes RN and Elner RW (1979) Tactics of a predator, *Carcinus maenas*, and morphological responses of the prey, *Nucella lapillus*. Journal of Animal Ecology 48: 65–78

Kitchell JF, Schindler DE, OgutuOhwayo R and Reinthal PN (1997) The Nile perch in Lake Victoria: interactions between predation and fisheries. Ecological Applications 7(2): 653–664

Kolar CS and Lodge DM (2001) Progress in invasion biology: predicting invaders. Trends in Ecology and Evolution 16(4): 199–204

Leonard GH, Bertness MD and Yund PO (1999) Crab predation, waterborne cues, and inducible defenses in the blue mussel, *Mytilus edulis*. Ecology 80: 1–14

Levinton JS (1982) Marine Ecology. Prentice-Hall Inc., New Jersey

Lewis DE and Cerrato RM (1997) Growth uncoupling and the relationship between shell growth and metabolism in the soft shell clam *Mya arenaria*. Marine Ecology Progress Series 158: 177–189

Lodge DM (1993) Biological invasions: lessons for ecology. Trends in Ecology and Evolution 8: 133–137

Maaski H and Guillou J (1999) The role of biotic interactions in juvenile mortality of the cockle (*Cerastoderma edule* L.): field observations and experiment. Journal of Shellfish Research 18: 575–578

McDonald PS, Jensen GC and Armstrong DA (2001) The competitive and predatory impacts of the nonindigenous crab *Carcinus maenas* (L.) on early benthic phase Dungeness crab *Cancer magister* Dana. Journal of Experimental Marine Biology and Ecology 258: 39–54

Meire PM and Ervynck A (1986) Are oystercatchers (*Haemoptopus ostralegus*) selecting the most profitable mussels (*Mytilus edulis*)? Animal Behavior 34: 1427–1435

Moody KE and Steneck RS (1993) Mechanisms of predation among large decapod crustaceans of the Gulf of Maine Coast: functional vs. phylogenetic patterns. Journal of Experimental Marine Biology and Ecology 168: 111–124

Nilsson PA (2001) Predator behaviour and prey density: evaluating density-dependent intraspecific interactions on predator functional responses. Journal of Animal Ecology 70: 14–19

Olson MH (1996) Predator–prey interactions in size-structured fish communities: implications of prey growth. Oecologia 108(4): 757–763

Peacor SD and Werner EE (2001) The contribution of trait-mediated indirect effects to the net effects of a predator. Proceedings of the National Academy of Sciences 98: 3904–3908

Rangeley RW and Thomas MLH (1987) Predatory behavior of juvenile shore crab *Carcinus maenas* (L.). Journal of Experimental Marine Biology and Ecology 108: 191–197

Relyea RA (2001) Morphological and behavioral plasticity of larval anurans in response to different predators. Ecology 82(2): 523–540

Rejmanek M (2000) Invasive plants: approaches and predictions. Austral Ecology 25: 497–506

Ruiz GM, Carlton JT, Grosholz ED and Hines AH (1997) Global invasions of marine and estuarine habitats by non-indigenous species: mechanisms, extent, and consequences. American Zoologist 37: 621–632

Skilleter GA (1994) Refuges from predation and the persistence of estuarine clam populations. Marine Ecology Progress Series 109: 29–42

Smith TE, Ydenberg RC and Elner RW (1999) Foraging behaviour of and excavating predator, the red rock crab (*Cancer productus* Randall) on soft shell clam (*Mya arenaria* L.) Journal of Experimental Marine Biology and Ecology 238: 185–197

Stachowicz JJ and Hay ME (1999) Mutualism and coral persistence: the role of herbivore resistance to algal chemical defense. Ecology 80(6): 2085–2101

Stephens DW and Krebs JR (1986) Foraging Theory. Princeton University Press, Princeton, New Jersey

Tollrian R (1995) Predator-induced morphological defenses – costs, life-history shifts, and maternal effects in *Daphnia pulex*. Ecology 76(6): 1691–1705

Tollrian R and Dodson SI (1999) Inducible defenses in cladocera: constraints, costs, and multipredator environments. In: Tollrian R and Harvell C (eds) The Ecology and Evolution of Inducible Defenses, pp 177–202. Princeton University Press, Princeton, New Jersey

Trussel GC (1996) Phenotypic plasticity in an intertidal snail: the role of a common crab predator. Evolution 50: 448–454

Trussell GC and Smith LK (2000) Induced defenses in response to an invading crb predator: an explanation of historical and geographic phenotypic change. Proceedings of the National Academy of Sciences, USA 97(5): 2123–2127

Van Dover C and Kirby-Smith WW (1979) Field Guide to Common Marine Invertebrates of Beaufort, NC. Duke University Marine Laboratory, Beaufort, North Carolina, 80 pp

van der Veer HW, Feller RJ, Weber A and Witte JIJ (1998) Importance of predation by crustaceans upon bivalve spat in the intertidal zone of the Dutch Wadden Sea as revealed by immunological assays of gut contents. Journal of Experimental Marine Biology and Ecology 231: 139–157

Virnstein RW (1977) The importance of predation by crabs and fishes on benthic infauna in Chesapeake Bay. Ecology 58: 119–1217

Welch WR (1968) Changes in abundance of the green crab, *Carcinus maenas* (L.), in relation to recent temperature changes. Fisheries Bulletin of the US Fish and Wildlife Service 67(2): 337–345

Whitlow WL (2002) Changes in native species after biological invasion: effects of introduced green crabs on native soft-shell clams and an estuarine community. PhD thesis, University of Michigan, Ann Arbor, Michigan

Wilson WH (1991) Competition and predation in marine soft-sediment communities. Annual Review of Ecology and Systematics 21: 221–241

Witte F, Goldschmidt T, Wanink J, Vanoijen M, Goudswaard K, Wittemaas E and Bouton N (1992) The destruction of an endemic species flock – quantitative data on the decline of the Haplochromine cichlids of Lake Victoria. Environmental Biology of Fishes 34(1): 1–28

Zaklan SD and Ydenberg R (1997) The body size – burial depth relationship in the infaunal clam *Mya arenaria*. Journal of Experimental Marine Biology and Ecology 215: 1–17

Biological Invasions **5:** 33–43, 2003.

Biotic resistance experienced by an invasive crustacean in a temperate estuary

Christopher E. Hunt[1,3] & Sylvia BehrensYamada[2,*]
[1]*Environmental Science Program,* [2]*Zoology Department, Oregon State University, Corvallis, OR 97331-2914, USA;* [3]*Scientific Applications International Corporation, 18706 N. Creek Parkway, Suite 110, Bothell, WA 98011, USA;* **Author for correspondence (e-mail: yamadas@science.oregonstate.edu; fax: +1-541-737-0501)*

Received 30 June 2001; accepted in revised form 11 March 2002

Key words: biotic resistance, *Cancer productus*, *Carcinus maenas*, green crab, invasion, predation, temperate estuary

Abstract

Communities high in species diversity tend to be more successful in resisting invaders than those low in species diversity. It has been proposed that the biotic resistance offered by native predators, competitors and disease organisms plays a role. In Yaquina Bay, Oregon, we observed very little overlap in the distribution of the invasive European green crab, *Carcinus maenas*, and the larger red rock crab, *Cancer productus*. *C. productus* dominates the more saline, cooler lower estuary and *C. maenas*, the less saline, warmer upper estuary. Because caged *C. maenas* survive well in the lower estuary, we decided to test the hypothesis that *C. productus* prey on *C. maenas* and thus contribute to their exclusion from the more physically benign lower estuary. A laboratory species interaction experiment was designed to determine whether *C. productus* preys on smaller *C. maenas* at a higher rate than on smaller crabs of their own species. Crabs of both species were collected and sorted by weight into three size classes: small, medium and large. Small and medium crabs of both species were paired with *C. maenas* and *C. productus* of various sizes. When conspecifics were paired, mortality was less than 14%, even in the presence of larger crabs. Smaller *C. productus* survived well in the presence of larger *C. maenas*, but the reverse was not true. When small *C. maenas* (60–67 mm carapace width) were matched with medium and large *C. productus*, their mortality increased to 52% and 76%, respectively. A less dramatic pattern was observed for medium *C. maenas* (73–80 mm) in the presence of medium and large *C. productus*. Thus on the West Coast of North America, the more aggressive red rock crab, *C. productus*, has the potential to reduce the abundance of *C. maenas* in the more saline and cooler lower estuaries.

Introduction

One of the most difficult questions facing invasion biologists is: why do exotic species invade some communities and not others? It has been proposed that islands and disturbed habitats are especially prone to invasions because the species-poor native communities lack predators, competitors, and disease organisms and thus offer little biotic resistance (Stachowicz et al. 1999). The success of goats on oceanic islands and the establishment of the Asian clam *Potamocorbula amurensis* in disturbed temperate estuaries are two examples that support this view (Coblentz 1978; Nichols et al. 1990; Cohen and Carlton 1998).

It may be possible to test the biotic resistance hypothesis, by comparing the success of a global invader in ecosystems differing in species diversity. The European green crab, *Carcinus maenas*, with its long invasion history is a good candidate for such a comparison. This species became established on the Atlantic coast of North America and the southeast coast of Australia prior to 1900 and recently appeared in

South Africa, Tasmania and on the Pacific coast of North America (Say 1817; Fulton and Grant 1900; Le Roux et al. 1990; Cohen et al. 1995; Grosholz and Ruiz 1995). Like many invasive species, *C. maenas* is a physiological and ecological generalist, tolerating air exposure, starvation and wide ranges in temperature and salinity (Behrens Yamada 2001). It consumes prey from over 150 genera including molluscs, crustaceans, worms, algae and marsh vegetation (Cohen et al. 1995). In its native habitat and on the Atlantic coast of North America, *C. maenas* inhabits estuarine tidal marshes, mudflats, cobble beaches, as well as rocky shores on both wave-protected and semi-exposed shores (Crothers 1970; Menge 1983; Grosholz and Ruiz 1996). When *C. maenas* is abundant, young mussels, urchins, cockle beds and barnacles cannot establish themselves (Kitching et al. 1959; Muntz et al. 1965; Jensen and Jensen 1985; Beukema 1991; Leonard et al. 1999). On the Atlantic coast of North America, the build up of *C. maenas* densities (up to 500/m^2) has been correlated with a dramatic decline in the landings of the soft-shelled clam, *Mya arenaria* (Glude 1955; MacPhail et al. 1955; Welch 1968).

While *C. maenas* is considered a pest on the Atlantic coast of North America, its presence in southeastern Australia causes little concern. Natural enemies, including predators, competitors and parasites, appear to keep densities low and impact small. Native Australian predators include the heavily armored crab *Ozius truncatus*, the aggressive swimming crab *Portunus pelagicus,* skates and rays (Zeidler 1997; Center for Research on Introduced Marine Pests information sheet). Sinclair (1997) found that seven species of native grapsid crabs were not adversely affected by the invader. While we lack information on the early invasion process of *C. maenas* in Australia, it appears that the resistance offered by the diverse native fauna, including other crabs, may prevent *C. maenas* from becoming a pest.

If the biotic resistance hypothesis applies to *C. maenas*, we might expect this invader to exert less of an impact on marine communities on the Pacific coast of North America than on the Atlantic coast. Jamieson et al. (1998) points out that 23 native brachyuran crabs species occur on the Pacific compared to only nine on the Atlantic coast. In our surveys of Oregon estuaries we typically find 10 native crabs within the distributional range of *C. maenas*. The most common are the grapsids, *Hemigrapsus oregonensis* (Oregon shore crab) and *H. nudus* (purple shore crab) and the cancrids, *Cancer productus* (red rock crab) and juvenile *Cancer magister* (Dungeness crab). Both species of *Hemigrapsus* are small omnivorous crabs that rarely attain a carapace width of over 30 mm. They possess relatively weak monomorphic claws with fine denticles and mechanical advantages of 0.28. (Mechanical advantage is the ratio of a claw's lever arms: the distance from the fulcrum to the attachment of the closer muscle and the distance from the fulcrum to the tip of the moveable finger.) The two species of *Cancer* are much larger, attaining carapace widths of over 160 mm for *C. productus* and over 190 mm for *C. magister*. While *C. magister* attains a larger size, its slender claws exhibit a mechanical advantage of only 0.25. *C. productus*, on the other hand, has very powerful claws for its size with a mechanical advantage of 0.39 (Behrens Yamada and Boulding 1998). *C. maenas*, with a carapace width of over 90 mm, is a medium sized crab compared to *Hemigrapsus* spp. and *Cancer* spp. Adult *C. maenas* possess claws of different sizes. The larger crusher claw has a mechanical advantage of 0.36 while the more slender cutter claw, 0.26 (Warner et al. 1982). This dual tool set may allow *C. maenas* to exploit a larger food spectrum than *Hemigrapsus* spp. and *Cancer* spp. with their monomorphic claws.

While surveying Oregon estuaries for the status of the *C. maenas* invasion during the summer of 1998, we noted very little overlap in the distribution of *C. maenas* and *C. productus*. *C. maenas* was the dominant crab in the warmer, less saline upper estuary and *C. productus*, in the cooler, more saline low estuary. A similar pattern was observed in Bodega Bay Harbor, California. *C. maenas* was restricted to the shallow mudflats inside the harbor while the brown rock crab, *Cancer antennarius*, and *C. productus* occupied the rocky shore close to the mouth of the harbor (Grosholz and Ruiz 1996; McDonald et al. 1998). When five *C. maenas* were caged on a rocky shore near the mouth of an Oregon estuary, they survived well (authors' unpublished data). This observation suggests that physical factors are not responsible for keeping *C. maenas* out of the lower estuary. Since *C. productus* has a reputation as a voracious predator on mollusks and other crabs, including smaller members of their own species (Daley 1981; Boulding and Hay 1984; Robles et al. 1989; Behrens Yamada and Boulding 1996), we suspected that *C. productus* predation may limit the distribution of *C. maenas* to the upper estuary. The goal of our study was twofold: (1) to document the distributional relationships between *C. maenas* and common native crabs, and (2) to test the hypothesis that predation by the native *C. productus* limits the distribution and abundance of nonindigenous *C. maenas*. These objectives were accomplished by systematically trapping crabs along a temperate Oregon estuary and by setting up laboratory arenas to elucidate the interactions between *C. maenas* and *C. productus* of various sizes.

Materials and methods

Distribution study

In order to document the distributional patterns between *C. maenas* and common native crabs, we sampled 10 stations, representing a range of habitat types, along the Yaquina estuary (Figure 1, Table 1). Since vandalism of traps was a problem, we focused our efforts on five permanent sites that were relatively secure from public interference: Port of Toledo docks, Riverbend Marine on Oneatta Point, Roberts' private dock by Weiser Point, a mudflat near the Oregon Coast Aquarium (OCA) and South Beach Marina. These sites were trapped daily for 90 days from June to September 1999 while five less secure sites were trapped for only 18 days.

All but one of the study sites possess rock shelters in the form of rip-rap close to the trapping sites. The OCA site, a tidal flat intersected by a channel, lacks rock cover. Temperature and salinity measurements were taken whenever traps were checked. Air and water temperatures were measured with a field thermometer and salinity of the surface water, with a temperature-compensated refractometer. Tidal levels of the traps were estimated from tide tables using the observed low tide mark as a reference point.

Two trap designs were used to collect a range of crab sizes. The collapsible Fukui fish trap (60 cm × 45 cm × 20 cm) is covered with 2 cm plastic mesh and possesses two expandable slit openings (45 cm) at each end. This design allows even the largest crabs to enter. The box trap is a modified 2 cm mesh stainless steel cage (60 cm × 60 cm × 30 cm) with conical meshed openings (8–9 cm in diameter) on each of the four sides. A commercial bait container, filled with fresh salmon backbone and flesh, was added to each trap each day. Two to three traps of each type were used at each of the sites. Traps were placed near the 0 m tidal level, except at the Sally's Bend mudflat where traps were set at the 0.5 m level. Traps were allowed to soak for a

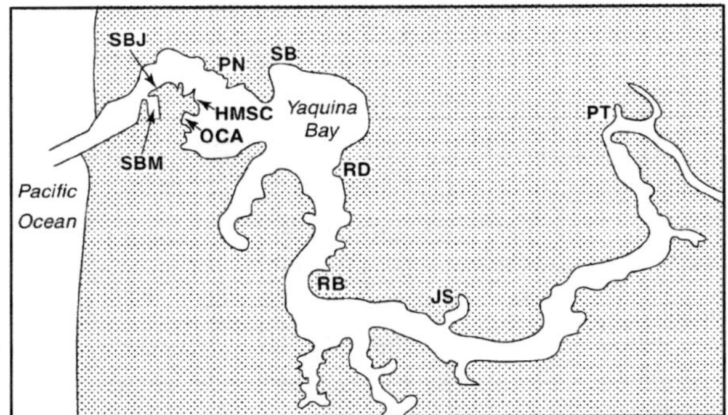

Figure 1. Map of Yaquina Bay, Oregon, showing locations of the ten trapping sites. For abbreviations and characteristics of sites see Table 1.

Table 1. Description of the physical features of each of the 10 trapping sites. Permanent sites with more than 208 trap days each are indicated by an asterisk.

Site	River mile	Salinity range (ppt)	Temperature range (°C)	Estimated slope	Habitat type
Port of Toledo, PT*	12.5	14–24	19–21	20–70°	Upper intertidal *Scirpus*, adjacent to rip-rap.
Johnson Slough, JS	7.4	16–28	15–18	10–30°	Upper intertidal *Scirpus*, adjacent to rip-rap.
Riverbend, RB*	5.5	21–35	14–16	30–50°	Rip-rap finger jetty, adjacent to river channel and tidal flat.
Roberts' Dock, RD*	4.6	28–36	11–19	10–45°	Intertidal *Zostera* on tidal flat, adjacent to rip-rap.
Sally's Bend, SB	3.1	15–30	13–20	<10°	Large tidal flat with upper intertidal *Scirpus*.
Oregon Coast Aquarium, OCA*	2.2	28–36	14–20	<10°	Upper intertidal *Scirpus*, minor rip-rap. Sea-water from aquarium outfall constantly drains into a channel that intersects a mudflat.
Port of Newport, PN	1.9	25–32	12–15	<10–45°	Rip-rapped shoreline adjacent to tidal flat.
Pumphouse, HMSC	1.8	23–30	12–14	10–45°	Rip-rapped finger jetty entering lower intertidal *Zostera* and sandy subtidal.
South Beach Marina Jetty, SBJ	1.3	28–34	12–16	10–45°	Rip-rapped shoreline adjacent to intertidal *Zostera* tidal flat.
South Beach Marina, SBM*	1.3	22–36	13–16	<10–45°	Rip-rapped shoreline entering subtidal sand and fine sediments.

day before they were checked. All trapped crabs were identified to species and sex and any injuries or parasite infections noted. Maximum carapace width, including protruding spines, was measured using vernier calipers. For example, *C. maenas* was measured between the tips of their 5th anterio-lateral spines and *C. productus,* between the tips of their 10th anterior-lateral spine. Catch per unit effort (CPUE) was calculated for each site by averaging the number of crabs caught per trap per day.

Laboratory predation experiment

Since we observed an inverse distributional relationship between *C. maenas* and *C. productus*, we set up pair-wise laboratory trials to determine whether the native *C. productus* preys on smaller *C. maenas* at a greater rate than on similar sized crabs of its own species. In order to set up a fair interaction trial between two crabs, we used only healthy crabs that had not molted recently or were not about to molt soon. We collected healthy male crabs of both species with intact limbs, weighed and measured them and derived regression equations of mass on carapace width:

$$C.\ maenas \quad \mathrm{Log}_{10}(\mathrm{Mass}) = \mathrm{Log}_{10}(3.12\mathrm{CW} - 3.83) \quad R^2 = 0.96$$

$$C.\ productus \quad \mathrm{Log}_{10}(\mathrm{Mass}) = \mathrm{Log}_{10}(2.93\mathrm{CW} - 3.73) \quad R^2 = 0.98$$

These regressions allowed us to identify animals that did not fall within the predicted 95% confidence interval of the regressions. Thus, crabs that were about to molt soon (heavy for their CW) or had recently molted (light for their CW) could be discarded. Three distinct weight classes of crabs were chosen: small (52–78 g), medium (85–125 g) and large (166–214 g). *C. maenas* from the abundant 1997/1998 year class were divided into small (60–67 mm CW) and medium (73–80 mm CW) size classes. Since *C. productus* reaches a maximum size twice that of *C. maenas*, three size classes were chosen: small (71–77 mm), medium (91–98 mm) and large (103–109 mm). Although still not approaching the maximum size (>190 mm) for *C. productus*, the large size class (103–109 mm) was used to determine if size plays a factor in the species interactions.

Upon entering the laboratory each healthy male crab was put into a conspecific community tank (120 cm × 43 cm × 30 cm) and fed mussels. Three days prior to starting an experiment, crabs were chosen at random, weighed and placed into individually numbered containers and starved. This allowed us to standardize hunger level and to randomly pair crabs from appropriate size classes. Small and medium *C. maenas* and small, medium and large *C. productus* were each paired with *C. maenas* and *C. productus* of various sizes. We did not run the interaction between two large *C. productus* as this interaction was not the focus of our study. Each of the 14 pairings was replicated 50 times (Table 2).

Table 2. Experimental design for the *C. maenas–C. productus* species interaction experiment. S, M and L refers to small medium and large crabs. Each of the 14 pairings was run 50 times.

	Carcinus (S)	*Carcinus* (M)	*Cancer* (S)	*Cancer* (M)
Carcinus (S)	50			
Carcinus (M)	50	50		
Cancer (S)	50	50	50	
Cancer (M)	50	50	50	50
Cancer (L)	50	50	50	50

Experimental arenas consisted of 3.8 liter rectangular plastic Sterilite® boxes (15.8 cm × 30 mm × 8 cm) with secured plastic lids. Filtered seawater (minimum flow of 2 l/h) entered via the top and exited via six out-flow holes along the sides. Thus, each arena had its own water source. No sediment was added to the arenas in an attempt to avoid possible variation due to sediment type. The salinity ranged from 32 to 37 ppt, and temperature from 11 °C to 14 °C. Each pair of crabs was placed simultaneously at opposite ends of an arena (Figure 2). Seven arenas were placed into each of eight fully drained outdoor aquariums with a semi-opaque lids and sides. Four replicates of each of the 14 treatments were run concurrently to minimize variation through time (day length, salinity, and temperature). Each day, for seven days, crabs were checked for limb loss and predation events.

Results

Distribution study

The ten study sites in Yaquina Bay reflect the typical habitat diversity and physical factors found in temperate estuaries (Figure 1, Table 1). As expected, the Port of Toledo in the upper estuary exhibited the lowest salinities and highest temperatures while sites near

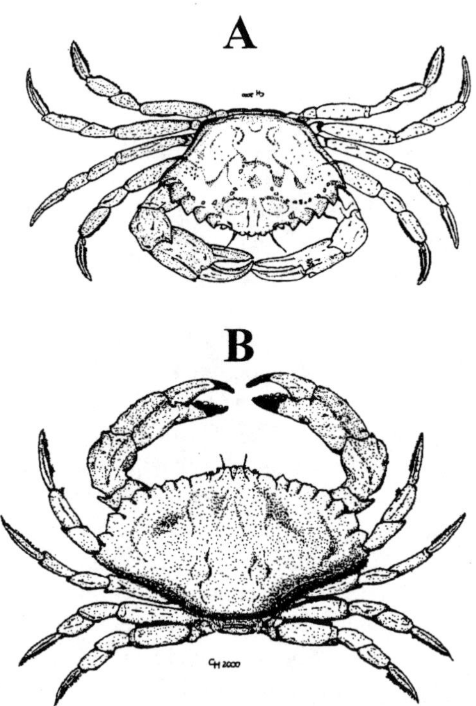

Figure 2. Top view of an experimental arena showing a medium *C. maenas* (A) paired with a medium *C. productus* (B).

the mouth of the estuary, the highest salinities and lowest temperatures. Sally's Bend with its extensive shallow mudflat exhibited salinities as low as 15 ppt and temperatures as high as 20 °C. The greatest variability in temperature was observed in the mid-estuary, reflecting the movement of upper and lower estuarine waters during the tidal cycle.

While crabs were trapped at all ten study sites, species diversity increased toward the cooler, high saline lower estuary (Table 3). *C. maenas* and the native crabs *C. productus*, juvenile *C. magister* and *H. oregonensis* were the most frequently trapped crabs. The native crabs *H. nudus*, *C. antennarius* and *Pugettia producta* occasionally entered traps in the lower estuary. The euryhaline *H. oregonensis* occurred throughout the estuary while the nonindigenous brackish water crab, *Rhithropanopeus harrisii* was only trapped at the Port of Toledo.

While crab species showed different preferences for the two trap types, the distributional patterns for the crabs remained the same. *C. maenas* entered box traps twice as often as the Fukui fish traps, while *C. magister* exhibited the opposite preference (Table 3). *H. oregonensis* was present at all 10 sites. The juveniles

Table 3. CPUE for the four most frequently caught crabs at the ten sites using Fukui fish traps (A) and box traps (B). CPUE is given as number of crabs per trap per day.

Site	*Carcinus maenas*	*Cancer productus*	*Cancer magister*	*Hemigrapsus oregonensis*
A. Fukui fish trap CPUE				
Port of Toledo, PT*	–	–	–	0.204
Johnson Slough, JS	0.869	–	1.986	0.331
Riverbend, RB*	0.742	0.017	0.922	0.104
Roberts Dock, RD*	0.551	0.008	1.884	1.045
Sally's Bend, SB	0.5780	–	–	0.154
Aquarium, OCA*	0.813	0.066	0.206	2.180
Port of Newport, PN	–	0.958	0.152	0.030
Pumphouse, HMSC	0.269	1.418	1.038	0.090
South Beach Jetty, SBJ	0.159	3.347	1.068	–
South Beach Marina, SBM*	–	5.857	0.222	0.148
B. Box trap CPUE				
Port of Toledo, PT*	–	–	–	0.056
Johnson Slough, JS	1.517	–	0.032	0.570
Riverbend, RB*	2.809	0.009	0.080	0.044
Roberts Dock, RD*	1.337	–	0.050	0.350
Sally's Bend, SB	1.050	–	–	0.011
Aquarium, OCA*	2.337	–	0.131	4.279
Port of Newport, PN	–	1.549	–	0.056
Pumphouse, HMSC	0.024	2.083	0.057	0.328
South Beach Jetty, SBJ	0.625	2.133	0.010	0.084
South Beach Marina, SBM*	–	4.477	0.099	–

of the commercial Dungeness crab, *C. magister*, ranged throughout the estuary, but were absent from the two sites where salinities dropped to 15 ppt and temperatures rose to 20 °C. *C. maenas* was most abundant between River Mile 2 and 7.4 but was noticeably absent from the Port of Toldedo were salinities were the lowest and from the Port of Newport and South Beach Marina near the mouth of the estuary. *C. productus* was

Table 4. Correlation of CPUE of *C. maenas* with three of the most common native crabs using Fukui fish traps (A) and box traps (B). The Pearson correlation coefficient is based on nine study sites. The Port of Toledo was eliminated from the analysis, as salinities there are too low to support most crab species.

Green crab vs. native species	Pearson correlation coefficient	Probability
A. Fukui fish trap		
C. maenas vs. *C. productus*	−0.747	0.021
C. maenas vs. *C. magister*	0.368	0.330
C. maenas vs. *H. oregonensis*	0.515	0.156
B. Box trap		
C. maenas vs. *C. productus*	−0.746	0.021
C. maenas vs. *C. magister*	0.421	0.259
C. maenas vs. *H. oregonensis*	0.471	0.200

the most abundant crab in the lower estuary from River Mile 1–2.

Correlations of *C. maenas* catches with those of each of the three common native species indicates no significant correlation with juvenile *C. magister* and *H. oregonensis* but a significant negative correlation in the distribution pattern with *C. productus* (Table 4). This negative correlation is observed with both trap types and is particularly dramatic for the five permanent study sites (Figure 3).

Laboratory predation experiment

Theoretically, the pairing of any two crabs could result in a predation event occurring in either direction. When conspecific crabs were paired, mortality was less than 14%, even in the presence of larger crabs (Figure 4). Smaller *C. productus* survived equally well in the presence of larger green crabs, but the reverse was not true. When small *C. maenas* were matched with medium *C. productus*, they experienced a 52% mortality. When matched with large *C. productus*, mortality increases even further to 76%. A similar pattern was found when medium *C. maenas* were matched with medium and large *C. productus*. Mortality increased to 32% and 46%, respectively. Chi-squared tests on the effect of *C. productus* size on the survival of *C. maenas* were highly significant (small *C. maenas*: $\chi^2 = 42.00$, df = 2, $P =< 0.0005$; medium *C. maenas*: $\chi^2 = 25.84$, df = 2, $P =< 0.0005$). Thus, the larger the asymmetry in size in favor of *C. productus*, the greater the mortality experienced by *C. maenas*.

Discussion

Distributions

Crab species diversity along a temperate Oregon estuary decreased with distance from the mouth. In Yaquina Bay, we observed the native crabs *H. oregonensis*, *H. nudus*, *Pachygrapsus crassipes*, *Lophopanopeus bellus*, *Cancer oregonensis*, *C. magister*, *C. productus*, *C. antennarius*, *Cancer gracilis* and *P. producta* in the saline, cooler lower estuary, but only *H. oregonensis* and the nonindigenous brackish water specialist, *R. harrisii* at the Port of Toledo at River Mile 12.5. The distance crabs penetrate up an estuary is dictated by their physiological tolerances. For example, *H. oregonensis* is the most tolerant of the native crabs to sedimentation, and to a wide range of temperatures and salinities (Dehnel and Carefoot 1965; Daley 1981) and consequently thrives at all sites along the estuary from South Beach Marina at River Mile 1.3 to the Port of Toledo at River Mile 12.5. Juvenile *C. magister* and adult *C. maenas* appear to be slightly less tolerant in that they penetrate the estuary up to Johnson Slough at River Mile 7.4. In contrast, *C. productus,* with its narrow physiological tolerances, is only found in the lower estuary where salinities remain above 25 ppt, temperatures below 15 °C and fine sediment is absent (Daley 1981). This species thrives on gently sloping shores with boulder shelters but is absent from tidal marshes and mudflats.

While marginal physical factors set the distributional limit of crab species in the upper estuary, species interactions become more important in the physically benign lower estuary. Although *C. maenas* is physiologically able to tolerate the habitat type and physical conditions in the lower estuary, its densities decrease near the mouth. *C. maenas* is present at South Beach Jetty but not inside South Beach Marina where *C. productus* is most abundant. The strong inverse correlation in catches of *C. maenas* and *C. productus* suggest that large male *C. productus* prey on smaller *C. maenas*. Large male *C. productus* are voracious predators that prey not only on molluscs but also on other crabs, including their own species (Daley 1981). Robles et al. (1989) have shown that female and smaller male *C. productus* do not leave their shelters when large male *C. productus* are foraging. Our unpublished observations suggest that small *C. productus* will not enter a trap if a large male *C. productus* is

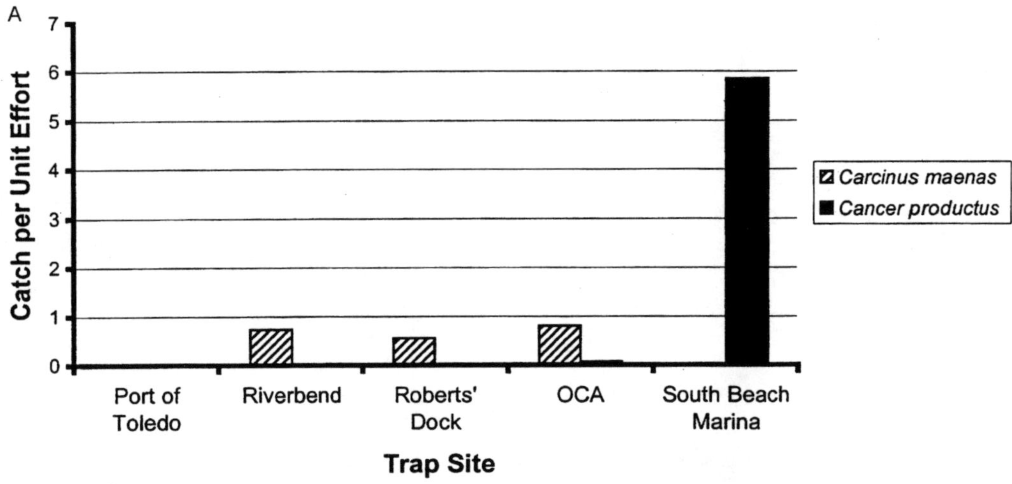

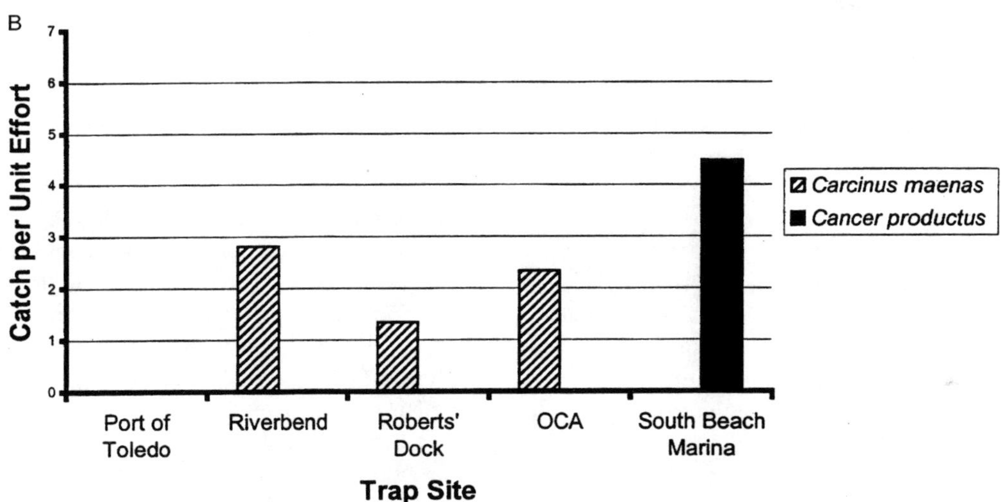

Figure 3. Primary site comparison: CPUE of *C. maenas* and *C. productus* at the five permanent trapping sites using Fukui fish traps (A) and box traps (B).

already present. It is conceivable that *C. maenas* exhibits similar avoidance behavior to the presence of larger *C. productus*. Thus, physiological tolerances prevent *C. productus* from exploiting the upper estuary while intraguild interactions play a role in excluding *C. maenas* from the more benign lower estuary.

Species interactions

Both species of crabs exhibited a low rate of cannibalism (14%) when confined to the arenas for seven days. Cannibalism upon the young by older members of the population is known to occur at some stage in the life history of many animal species (Polis 1981), and is very common among crabs, including *C. maenas* and *C. productus* (Elner 1981; Kurihara and Okamoto 1987; Anger 1995; Beck 1997; Lovrich and Sainte-Marie 1997; Moksnes et al. 1997). Cannibalism can regulate population size, reduce intra-specific competition, and supplement the diet with a nitrogen rich food source (Klein Breteler 1975; Kneib et al. 1999).

Size is an important consideration in any interaction between two crabs. Similar sized antagonists of the same species seldom kill each other (Huntingford et al. 1995; Lee 1995). Aggressive behavior directed against substantially smaller individuals, however, carries less

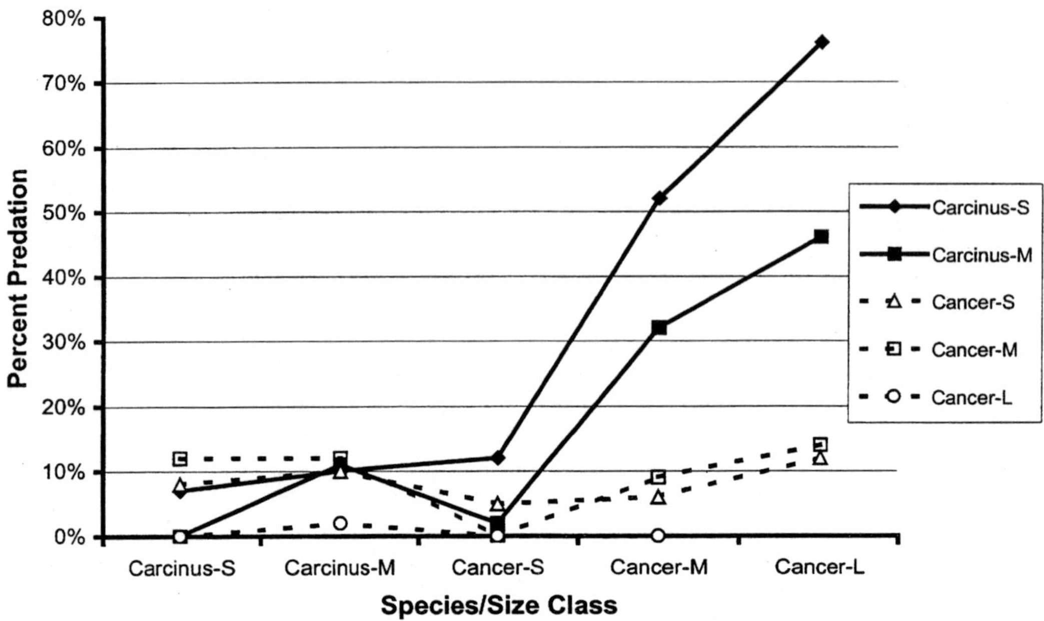

Figure 4. Percent predation experienced by crabs in the legend when paired with crabs on the *x*-axis. Cancer refers to *C. productus* and S, M and L to small medium and large crabs. See text for details.

risk of injury for the larger antagonist. Predation on conspecifics, however, did not increase with size in our arenas. Even small *C. productus* survived well when matched with large conspecifics. Small *C. productus* survived well in the presence of medium *C. maenas* but the reverse was not true. Larger *C. productus* preyed significantly more on smaller *C. maenas* than on conspecifics. The greater the size difference, the greater the predation rate. Small *C. maenas* experienced 12%, 52% and 76% predation respectively when matched with small, medium and large *C. productus*. Even medium *C. maenas* survived poorly when matched with similar sized *C. productus*. It appears that *C. maenas* with only one crusher claw with a mechanical advantage of 0.36 is no match for *C. productus* with two strong claws with a mechanical advantage of 0.39. Our laboratory trials demonstrate that direct predation by larger *C. productus* males can explain the low abundance or absence of *C. maenas* in our study sites near the mouth of the estuary.

Predictions

It is difficult to predict the eventual distribution, densities and impact of *C. maenas* on the Pacific shores of North America. With maximum densities of only 0.04–0.4/m^2 (Grosholz et al. 2000), this species has not reached the densities observed in its native range or on the Atlantic coast of North America. Currently, the species appears to be recruitment limited. The next El Niño event, however, could result in another density increase and range expansion (Behrens Yamada and Hunt 2000; Behrens Yamada 2001).

The results from our study suggests that at least one member of the native community, *C. productus*, is offering biotic resistance to the invasive *C. maenas* on the Pacific coast of North America. Direct predation, or the threat of predation, by larger *C. productus* appears to keep *C. maenas* out of habitats where dense populations of larger *C. productus* thrive: wave-protected rocky shores and gently sloping bolder beaches in cool, saline waters. On those beaches *C. maenas* larvae may settle high on the shore but as the young crabs mature and migrate down, they will encounter resistance from the larger *C. productus*. While large male *C. productus* spend most of their time in the subtidal, they do migrate into the intertidal with the tide to forage (Robles et al. 1989; Walker and Behrens Yamada 1993; Behrens Yamada and Boulding 1996). We predict that the native community on those beaches, where *C. productus* dominates, will not be heavily impacted by *C. maenas* and thus will retain much of their current species diversity.

The impact of *C. maenas* will be most intense in the estuaries and bays with soft sediment. Studies indicate that the native oyster (*Ostrea conchaphila*), mussels (*Mytilus trossulus*), thin shelled clams (e.g. *Nutricola* spp.) and *H. oregonensis* will be adversely affected (Bassett 2000; Grosholz et al. 2000; Palacios and Ferraro 2001). It is also feared that *C. maenas* predation and competition will reduce the abundance of the commercial Dungeness crab, *C. magister*, and flatfish that use estuarine mudflats as nursery habitat (Lafferty et al. 1996). Juvenile *C. magister* and small *H. oregonenesis* appear to offer little resistance to the more aggressive *C. maenas* in the upper estuary. Jensen et al. (2000) and McDonald et al. (2001) set up laboratory competition trials in which individual *C. maenas* were matched with *C. magister* or *H. oregonensis* of similar size. Crab pairs were introduced into arenas with either a food or oyster shells for shelters. *C. maenas* was a better competitor for food and shelter than similar sized *C. magister.* Claw size and mechanical advantage undoubtedly played a role in these interactions. Those of *C. magister* are much smaller and weaker than the crusher claw of similar sized *C. maenas* (Hauck 2000). While *C. maenas* won the food competition, *H. oregonensis* won the shelter competition (Jensen et al. 2000). Size for size adult *H. oregonensis* males have larger claws than juvenile *C. maenas* but young *C. maenas* have a higher metabolic rate and consume more food than *H. oregonensis* of similar size (S. Mahaffy, unpub. student report).

It is too early to predict the eventual densities, range and impact of *C. maenas* on the Pacific coast of North America. It is also too early to test the biotic resistance hypothesis that high crab species diversity on the Pacific coast will prevent *C. maenas* from becoming less of a pest than on the Atlantic coast where crab species diversity is low (Jamieson et al. 1998). The biotic resistance hypothesis, however, does apply on a smaller geographic scale. We observed that the species-poor community in the upper estuary is more prone to invasions by exotic crabs than the species rich community in the lower estuary. The brackish water crab, *R. harrisii*, from the Atlantic coast of North America was introduced to the Pacific coast in the 1930s, most likely with oyster transplants (Carlton 1979). It is now established in uppermost estuary of some California and Oregon estuaries where it encounters only one native species: the euryhaline *H. oregonensis*. Jordan (1989) found that the larger and more aggressive native *H. oregonensis* preys on juvenile *R. harrisii* and may force this invader to move into lower saline waters than in its native habitat. The invasive *C. maenas* thrives in the upper estuary where it coexists with *H. oregonensis* and juvenile *C. magister* but not in the species rich lower estuary where it encounter resistance from the aggressive *C. productus*.

Acknowledgements

This research was carried out by CEH in partial fulfillment for the Masters of Science degree in the OSU Environmental Science Program. We thank Bruce Coblentz, Jessie Ford, Laura Hauck, Eric Skyllingstad, Robert Olson, Kelly Palacios, and three anonymous reviewers for commenting on previous versions of this manuscript. Kelly Palacios helped check traps at the Oregon Coast Aquarium site and experimental arenas. This study was funded by the National Sea Grant College Program of the US Department of Commerce's National Oceanic and Atmospheric Administration under NOAA grant Number NA76RG0476 (Project number R/NIS-2-PD), and by appropriations made by the Oregon State legislature.

References

Anger K (1995) Developmental biology of *Armases miersii* (Grapsidae), a crab breeding in supratidal rock pools. II. Food limitation in the nursery habitat and larval cannibalism. Marine Ecology Progress Series 117: 83–89

Bassett Z (2000) The impact of the introduced green crab, *Carcinus maenas*, on selected bivalves and meiofauna communities. Master's thesis, Portland State University

Beck MW (1997) A test for the generality of the effects of shelter bottlenecks in four stone crab populations. Ecology 78: 2847–2503

Behrens Yamada S (2001) Global Invader: the European Green Crab. Oregon Sea Grant, Washington Sea Grant, Corvallis, Oregon, 123 pp

Behrens Yamada S and Hunt C (2000) The arrival and spread of the European green crab, *Carcinus maenas*, in the Pacific Northwest. Dreissena 11(2): 1–7

Behrens Yamada S and Boulding EG (1996) The role of highly mobile crab predators in the intertidal zonation of their gastropod prey. Journal of Experimental Marine Biology and Ecology 204: 59–83

Behrens Yamada S and Boulding EG (1998) Claw morphology, prey size selection and foraging efficiency in generalist and specialist shell-breaking crabs. Journal of Experimental Marine Biology and Ecology 220: 191–211

Beukema JJ (1991) The abundance of shore crabs *Carcinus maenas* (L.) on a tidal flat in the Wadden Sea after cold and mild winter. Journal of Experimental Marine Biology and Ecology 153: 97–113

Boulding EG and Hay TK (1984) Crab response to prey density can result in density-dependent mortality of clams. Canadian Journal of Fisheries and Aquatic Sciences 41: 421–525

Carlton JT (1979) History, biogeography, and ecology of the introduced marine and estuarine invertebrates of the Pacific coast of North America. PhD thesis, University of California at Davis

Coblentz BE (1978) The effect of feral goats (*Capra hircus*) on island ecosystems. Biological Conservation 13: 279–286

Cohen AN and Carlton JT (1998) Accelerating invasion rate in a highly invaded estuary. Science 279: 555–558

Cohen AN, Carlton JT and Fountain MC (1995) Introduction, dispersal and potential impacts of the green crab *Carcinus maenas* in San Francisco Bay, California. Marine Biology 122: 225–237

Crothers JH (1970) The distribution of crabs on rocky shores around the Dale Peninsula. Field Studies 3: 263–274

Daley GP (1981) Competitive interactions among three crab species in the intertidal zone. PhD thesis, University of Oregon, Eugene, Oregon

Dehnel PA and Carefoot TH (1965) Ion regulation in two species of intertidal crabs. Comparative Biochemistry and Physiology 15: 377–397

Elner RW (1981) Diet of green crab *Carcinus maenas* (L) from Port Hebert, Southwestern Nova Scotia. Journal of Shellfish Research 1(1): 89–94

Fulton SW and Grant FE (1900) Note on the occurrence of the European crab *Carcinus maenas*, Leach, in Port Phillip. Victorian Naturalist 27: 147–148

Glude JB (1955) The effects of temperature and predation on the abundance of soft-shell clam *Mya arenaria* in New England. Transactions of the American Fisheries Society 84: 13–26

Grosholz ED and Ruiz GM (1995) Spread and potential impact of the recently introduced European green crab, *Carcinus maenas*, in central California. Marine Biology 122: 239–247

Grosholz ED and Ruiz GM (1996) Predicting the impact of introduced marine species: lessons from the multiple invasions of the European green crab *Carcinus maenas*. Biological Conservation 78: 59–66

Grosholz ED, Ruiz GM, Dean CA, Shirley KA, Maron JC and Connors PG (2000) The implications of a nonindigenous marine predator in a California bay. Ecology 81: 1206–1224

Huntingford FA, Taylor AC, Smith IP and Thorpe KE (1995) Behavioural and physiological studies of aggression in swimming crabs. Journal of Experimental Marine Biology and Ecology 193: 21–39

Hauck L (2000) Use of tethered prey for estimating the impact of the invasive European green crab. Senior thesis, Oregon State University, Corvallis, Oregon

Jamieson GS, Grosholz ED, Armstrong DA and Elner RW (1998) Potential ecological implications from the introduction of the European green crab, *Carcinus maenas* (Linneaus), to British Columbia, Canada, and Washington, USA. Journal of Natural History 32: 1587–1598

Jensen KT and Jensen JN (1985) The importance of some epibenthic predators on the density of juvenile benthic macrofauna in the Danish Wadden Sea. Journal of Experimental Marine Biology and Ecology 89: 157–174

Jensen GC, McDonald PS and Armstrong DA (2000) East meets west: competitive interactions between green crab, *Carcinus maenas* and *Hemigrapsus* spp. Abstract in Journals of Shellfish Research 19(1): 157–174

Jordan JR (1989) Interspecific interactions between the introduced Atlantic crab *Rhithropanopeus harrisii* and the native estuarine crab *Hemigrapsus oregonensis* in Coos Bay, Oregon. MS thesis, University of Oregon, Eugene, Oregon

Kitching JA, Sloane JF and Ebling FJ (1959) The ecology of Louch Ine. VIII. Mussels and their predators. Journal of Animal Ecology 28: 331–3341

Klein Breteler WCM (1975) Laboratory experiments on the influence of environmental factors on the frequency of moulting and the increase in size at moulting of juvenile shore crabs, *Carcinus maenas*. Netherland Journal of Sea Research 1: 100–120

Kneib RT, Lee SY and Kneib JP (1999) Adult–juvenile interactions in the crabs *Sesarma* (Parisesarma) *bidens* and *S.* (Holometopus) *dehaani* (Decapoda: Grapsidae) from intertidal mangrove habitats in Hong Kong. Journal of Experimental Marine Biology and Ecology 234: 255–273

Kurihara Y and Okamoto K (1987) Cannibalism in a grapsid crab, *Hemigrapsus penicillatus*. Marine Ecology Progress Series 41: 123–127

Lafferty K and Kuris A (1996) Biological control of marine pests. Ecology 77: 1989–2000

Lee SY (1995) Cheliped size and structure; the evolution of the multi-functional decapod organ. Journal of Experimental Marine Biology and Ecology 193: 161–176

Leonard GH, Ewanchuk and Bertness MD (1999) How recruitment, intraspecific interaction, and predation control species borders in a tidal estuary. Oecologia 118: 492–502

LeRoux PJ, Branch GM and Joska MAP (1990) On the distribution, diet and possible impact of the invasive European shore crab *Carcinus maenas* (L) along the South African coast. South African Journal of Marine Science 9: 85–93

Lovrich GA and Sainte-Marie B (1997) Cannibalism in the snow crab, *Chionectes opilio* (*O. Fabricus*) (Brachyura: Majidae), and its potential importance to recruitment. Journal of Experimental Marine Biology and Ecology 211: 225–245

MacPhail JS, Lord EI and Dickie LM (1955) The green crab – a new clam enemy. Progress Report Atlantic Coast Stations 63: 3–12

McDonald PS, Jensen GC and Armstrong DA (2001) The competitive and predatory impacts of the nonindigenous crab *Carcinus maenas* (L) on early benthic phase Dungeness crab *Cancer magister* Dana. Journal of Experimental Marine Biology and Ecology 258(1): 39–54

Menge BA (1983) Components of predation intensity in the low zone of the New England rocky intertidal region. Oecologia 58: 141–155

Moksnes PO, Lipcius RN, Pihl L and van Montfrons J (1997) Cannibal-prey dynamics in young juveniles and postlarvae of the blue crab. Journal of Experimental Marine Biology and Ecology 215: 157–187

Muntz L, Ebling FJ and Kitching JA (1965) The Ecology of Lough Ine XIV. Predatory activity of large crabs. Journal of Animal Ecology 34: 315–329

Nichols FH, Thompson JK and Schemel LE (1990) Remarkable invasion of San Francisco Bay (California, USA) by the Asian clam *Potamocorbula amurensis*. II. Displacement of a former community. Marine Ecology Progress Series 66: 95–101 between a dominant competitor and its principal predator. Oecologia 15: 93–120

Palacios KC and Ferraro SP (2001) Food preference of the introduced green crab, *Carcinus maenas*, for the native Yaquina oyster, *Ostrea lurida*, compared to other local bivalves. Report prepared for the National Network for Environmental Management Studies. Abstract in Journal of Shellfish Research 19(1): 633

Polis GA (1981) The evolution and dynamics of intraspecific predation. Annual Review of Ecology and Systematics 12: 225–251

Robles C, Sweetnam DA and Dittman D (1989) Diel variation of intertidal foraging by *Cancer productus* L. In British Columbia. Journal of Natural History 23: 1041–1049

Say T (1817) An account of the Crustacea of the United States. Journal of the Academy of Natural Sciences, Philadelphia 1: 57–63

Sinclair MA (1997) Interactions between native grapsid crabs and *Carcinus maenas* in Victoria. Technical report 11. Center for Research on Introduced Marine Pests, Hobart, Australia

Stachowicz J, Whitlatch RB and Osman RW (1999) Species diversity and invasion resistance in a marine ecosystem. Science 286: 1577–1579

Walker S and Behrens Yamada S (1993) Implications for the gastropod fossil record of mistaken crab predation on empty mollusc shells. Palaeontology 36: 735–741

Warner GF, Chapman D, Hawkey N and Waring DG (1982) Structure and function of the chelae and chela closer muscles of the shore crab *Carcinus maenas* (Crustacea: Brachyura). Journal of Zoology, London 196: 431–438

Welch WR (1968) Changes in abundance of the green crab, *Carcinus maenas* (L.), in relation to recent temperature changes. Fishery Bulletin 67: 337–345

Zeidler W (1997) The European shore crab in South Australian waters. Technical report 11. Center for Research on Introduced Marine Pests, Hobart, Australia

Biological Invasions **5:** 45–51, 2003.

Alteration of microbial community composition and changes in decomposition associated with an invasive intertidal macrophyte

Daniel R. Hahn
Department of Zoology, University of Washington, Box 351800, Seattle, WA 98195-1800, USA (e-mail: hahnd@u.washington.edu; fax: +1-206-543-3041)

Received 15 July 2001; accepted in revised form 28 May 2002

Key words: Biolog EcoPlates, decomposition, eelgrass, invasions, microbial diversity, *Zostera japonica*, *Zostera marina*

Abstract

Data demonstrating the effects of biological invaders on microbial communities and microbial processes are scarce, especially in marine environments. Research was conducted at Padilla Bay, Washington, to examine the effect that an invasive intertidal eelgrass, *Zostera japonica* Aschers & Graebn, has on rates of decomposition, microbial community composition, and the possible implications for associated ecosystem processes in this estuarine environment. A series of observational and experimental studies were conducted in beds of *Z. japonica*, beds of its native congener, *Zostera marina*, and mixed eelgrass beds. These studies assessed decomposition of invasive and native eelgrass, enumerated bacterial abundance, and examined sole source carbon usage (SSCU) by microbial assemblages. *Z. japonica* decomposed more rapidly than its native congener throughout the study period although rates of decomposition were variable. Microbial abundance did not differ among different vegetation compositions although differences in SSCU by microbial assemblages were detected among beds of invasive, native, and mixed eelgrass. These results indicate that this abundant invasive species can accelerate rates of decomposition and alter the associative decomposer community, which may lead to higher carbon and nutrient turnover within Padilla Bay.

Abbreviation: SSCU – sole source carbon use

Introduction

Biological invasions often alter the community composition and dynamics of invaded areas (Posey 1988; Coles et al. 1999; MacIsaac et al. 1999) and can potentially alter ecosystem processes including productivity, decomposition, nutrient cycling, hydrology, and others (Elton 1958; Vitousek 1990; D'Antonio and Vitousek 1992; Vitousek et al. 1997). Quantitative data demonstrating such effects, however, are scarce, especially in marine environments (Ruiz et al. 1997, 1999). With anthropogenic activities introducing species outside of their natural range at an alarming rate (Lodge 1993; Cohen and Carlton 1995, 1998), understanding how invasive species may alter recipient communities and subsequent ecosystem properties is fundamental. One possible effect of invasive species is to modify the recipient community in such a way that subsequent changes in decomposition and/or nutrient cycling occur (Windham 2001). This may occur in one or both of the following ways: (1) providing organic material with a unique chemical composition and, (2) altering the decomposer community (primarily microbes) through either chemical or structural means. Because microbial processes are intricately linked to ecosystem level processes such as decomposition and nutrient cycling (Pedersen et al. 1999; Naeem et al. 2000), the consequences of an altered organic chemical regime or altered microbial community may be manifested as changes in the rate of decomposition or the ability of the native microbial community to degrade organic materials.

Although invading species can potentially change the composition of microbial communities, to my knowledge, no previous research has specifically addressed this topic. This may be due to the inherent difficulty of studying microbial communities. Until recently, assessing differences among microbial communities has been costly and/or very labor intensive (Garland 1997). With the development of new methods that allow a rapid, community-level assessment of microbes, comparisons among different sites have become much more tractable (Garland and Mills 1991; Ellis et al. 1995; Garland 1997; Stephan et al. 2000).

The goal of this research was to examine the effects that *Zostera japonica* Aschers & Graebn, has on microbial processes, rates of decomposition, and their potential influence on associated community and ecosystem properties in Padilla Bay, Washington. *Z. japonica,* also known as Asian eelgrass or dwarf eelgrass, was unintentionally introduced to Washington State with shipments of oysters from Japan. The date of introduction is not known but falls somewhere between the date of first collection, 50 years ago, and the first importation of Japanese oysters to the area, 100 years ago (Harrison 1976; Harrison and Bigley 1982). *Z. japonica* has since spread along the coastline, replacing unvegetated mudflat habitat (Posey 1988) and possibly competing with the native eelgrass, *Zostera marina* L. (Harrison 1982; Nomme and Harrison 1991a, b). *Z. japonica* tends to grow higher in the intertidal than its native congener although in many areas the two species overlap forming beds of mixed vegetation (Harrison 1982; Nomme and Harrison 1991a, b; Bulthuis 1995). Thus, in any comparisons between *Z. marina* and *Z. japonica*, tidal height is a confounding factor and must be parsed out using manipulative experiments. *Z. japonica* has been shown to have significant effects on infaunal assemblages and sediment structure (Posey 1988), but no exploration of its effect on microbial communities has been conducted. Presented here are the results of a combination of observational and experimental studies that measured rates of decomposition of native and introduced eelgrass, and quantified the difference in bacterial abundance and microbial community assemblages in beds of native, introduced, and mixed eelgrass while separating out the confounding effect of tidal height.

Study site and methods

This research was conducted at the Padilla Bay National Estuarine Research Reserve located in northern Puget Sound (latitude 48°28′ N, longitude 122°31′ W). Tidal fluctuations are quite large with a maximum range of nearly 4 m. With its predominance of intertidal flats, Padilla Bay supports one of the largest contiguous beds of *Z. marina* (native eelgrass) on the west coast of North America (Bulthuis 1995, for a more thorough site description and map see Bulthuis 1996). In addition to beds of the native eelgrass the bay also has extensive beds of the invasive *Z. japonica* as well as mixed eelgrass beds. Bulthuis (1991) mapped the eelgrass beds in 1989 finding ca. 2900 ha of *Z. marina*, 236 ha of *Z. japonica*, and 88 ha of mixed eelgrass. The coverage of both *Z. japonica* and mixed beds has increased substantially over the past decade (Hahn, pers. obs.). The zone of overlap between the native and invasive eelgrass is widest in the middle part of the bay and narrowest in the northern part of the bay (Hahn, unpub. data).

Eelgrass transplants

Within naturally occurring beds of eelgrass, the effects of bed composition and tidal height are confounded (Harrison 1982; Nomme and Harrison 1991a,b; Bulthuis 1991,1995). To separate the effects of these two factors, a series of transplants was conducted. These transplants moved the various bed compositions (*Z. japonica*, *Z. marina*, and mixed) to each tidal zone (upper or *Z. japonica* zone, lower or *Z. marina* zone, and the intermediate or mixed zone) with control beds transplanted back into the zone from which they originated. A total of nine transplants were established for this research and these transplants were conducted in the central portion of the bay where the mixed zone tends to be largest.

In the process of relocating each bed of eelgrass, turfs of eelgrass and sediment to a depth of ca. 10–15 cm were transplanted. This depth included the rhizomes and the majority of roots for both eelgrass species. Transplants measured 1 m^2 in size and were cut into 16 equal pieces to facilitate transport. Transplants were moved between sites using a plastic snow sled and each piece of sod was placed on a piece of plastic tarp to keep the sod together during handling. Once each transplant was in place, a 20 cm border was cleared around the plot to minimize the effect of surrounding vegetation on the transplant. Transplants were conducted in May of 2000 and beds were allowed to establish and grow for at least 4 weeks before any samples were taken.

Estimating decomposition rates

To determine decomposition rates, decomposition bags were placed in the field. In order to get accurate measures of the mass lost during experiments, eelgrass was collected fresh and oven dried at 40 °C. Each 8 cm × 10 cm bag was constructed of fine mesh (nylon sheer, 0.83 mm × 0.38 mm pore size) and contained 5 g (dry weight) of either *Z. japonica* or *Z. marina*. Bags were anchored to the sediment surface by attaching them to a loop in the end of a 30 cm piece of heavy gauge zinc wire. Within naturally occurring beds of eelgrass, 3 replicate bags of each vegetation type (18 bags total) were haphazardly deployed in early June 2000 and left in the field for 39 days. Within the transplanted beds, 3 replicate bags of each vegetation type were haphazardly placed in each plot (54 bags total) in mid June 2000 and left in the field for 27 days. Upon collection, all bags were rinsed thoroughly with tap water to remove sediments, dried at 40 °C, and reweighed.

Extraction of microbes for direct counts and SSCU

Microbial samples were obtained from the sediments by collecting soil cores in mid-July using a soil-coring device fitted with a 5 cm (internal diameter) by 15 cm aluminum sleeve. Within naturally occurring beds of eelgrass, cores were taken haphazardly with a total of five cores from *Z. japonica* beds (upper intertidal), eight cores from *Z. marina* beds (lower intertidal), and three cores from mixed eelgrass beds (intermediate tidal height). Within the transplanted eelgrass beds, three cores were haphazardly taken from each of the plots. Once a core was extracted, the sleeve was removed, capped at both ends, and kept refrigerated until processing the next day. In order to obtain workable levels of bacteria a dilution of the sediments was conducted. The top 5 cm of the core were homogenized and 5 g (wet weight) was sub-sampled from the homogenate. The 5 g of sediment was placed in a sterile 60 ml polyethylene centrifuge tube with 45 ml of filter-sterilized seawater, capped, and shaken vigorously for 30 min. One millilitre of the resulting suspension was removed from the centrifuge tube and transferred to a sterile 25 ml test tube to which an additional 9 ml of sterile seawater was added. At this point (1 : 100 dilution), 9 ml of the dilution was transferred to a scintillation vial and preserved with formalin for later use in assessing relative bacterial abundance. The remaining 1 ml was again diluted with 9 ml of sterile seawater for a final dilution of 1 : 1000 and was used in SSCU.

Assessing bacterial abundance

Formalin-preserved samples were stained with DAPI (4′,6-diamidino-2-phenylindole) at a concentration of 0.4 μg/ml and microbes were collected by running 0.1–0.5 ml of the sample through a 0.2 μm membrane filter (modified from Porter and Feig 1980; Bird et al. 2000). Filters were mounted on microscope slides, covered with immersion oil and a coverslip, and counted at 1000× using an epifluorescence microscope. Ten fields were counted per sample and these counts were averaged to give a single estimate of relative bacterial abundance for each sample.

SSCU

The final 1 : 1000 dilution, obtained from the sediment cores, was plated onto Biolog EcoPlates (Biolog Inc., Hayward, California) at a volume of 150 μl per well. Biolog EcoPlates contain 31 unique carbon sources (one sole carbon source per well) and a redox colorant (tetrazolium violet). When bacteria are able to use a carbon source for growth, the reduction of the colorant causes the well to turn purple and indicates that the inoculant microbial community is able to utilize that carbon source. Plates were incubated at room temperature and were read after 4, 12, 24, 48, 72, 96, and 120 h using a Biolog Micrlog Microstation Plate Reader. For consistency, data presented here are from the 72 h reading which showed a significant amount of color development.

Statistical analysis

Inferential statistics are applied to the data collected in this study to provide a basis for examining patterns among treatments (Oksanen 2001, but see Hurlbert 1984). Data from decomposition bags and from microbial enumeration were analyzed using ANOVA. Because many of the 31 measures of SSCU obtained from Biolog EcoPlates were highly correlated, these data were reduced to their principal components. Subsequently the first three principal components were subjected to MANOVA to test for effects of vegetation type and tidal zone on microbial SSCU. All analyses were conducted using SYSTAT 9.0 (SPSS Inc., 1999).

Results

Decomposition

Decomposition of eelgrass in the field showed a considerable amount of variation (Figure 1) with overall losses ranging from 50% to 80% during 39 days of incubation in naturally occurring beds and 38–60% during the 27-day deployment in transplanted beds. Within naturally occurring beds of eelgrass, bed composition/tidal zone did not have a significant effect on the rate at which either eelgrass species decomposed (Figure 1A, Table 1A). Similarly, within the transplanted eelgrass beds, neither the tidal height nor the composition of the eelgrass bed had a significant effect on the rate of decomposition (Figure 1B, Table 1B). In all cases, however, *Z. japonica* decomposed significantly faster than the native eelgrass (Figure 1, Table 1).

Bacterial enumeration

Direct counts of microbes using DAPI epifluorescence microscopy revealed no significant differences in the number of bacteria among vegetation types or tidal zones (Figure 2). This was true for bacteria isolated from sediments in naturally occurring beds of eelgrass (Figure 2A) as well as transplanted vegetation which separated out the effects of the bed composition from those of the tidal zone (Figure 2B).

SSCU

Multivariate data (results from 31 unique carbon sources) were reduced to their principal components and the first 3 principal components explained a total

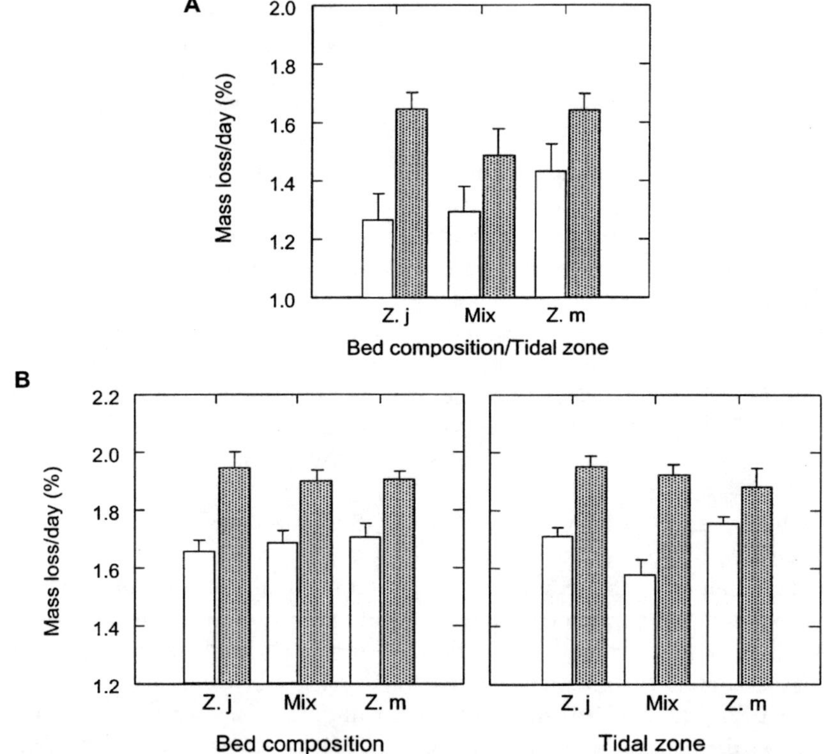

Figure 1. Decomposition of invasive eelgrass, *Z. japonica* (shaded bars) and native eelgrass, *Z. marina* (open bars) in eelgrass beds of different composition. A: Naturally occurring beds. B: Transplanted eelgrass beds. Values shown are mean ± SE. *Z.j* = *Zostera japonica*, Mix = mixed vegetation (*Z. japonica* and *Z. marina*), *Z.m* = *Z. marina*. See Table 1 for ANOVA results.

of 40% of the variance associated with the original 31 measures of SSCU. These first three principal components were used in the subsequent MANOVA to test for effects of bed composition, tidal zone, and any possible interaction between them (Table 2). Results of the MANOVA revealed that the bed composition had a significant effect on the microbial assemblage while tidal height did not have a significant effect. The interaction between vegetation and tidal zone was significant for the first principal component, but the multivariate test statistics was not significant.

Table 1. ANOVA results for decomposition of invasive and native eelgrass in eelgrass beds of different composition.

Source	SS	df	MS	*F*	*P*
A. Within naturally occurring beds					
Bed composition/ tidal zone	0.037	2	0.019	2.126	0.162
Vegetation type	0.240	1	0.240	27.475	<0.001
Interaction	0.026	2	0.013	1.494	0.263
Error	0.105	12	0.009		
B. Transplanted eelgrass beds					
Bed composition	0.022	2	0.011	0.039	0.961
Tidal zone	1.672	2	0.836	2.960	0.065
Vegetation type	19.051	1	19.051	67.466	<0.001
Bed comp. × T. zone	0.994	4	0.248	0.880	0.485
Bed comp. × Veg. type	0.510	2	0.255	0.903	0.415
T. zone × Veg. type	2.536	2	1.268	4.491	0.018
Bed comp. × T. Zone × Veg. type	0.256	4	0.064	0.227	0.921
Error	10.166	36	0.282		

Discussion

Over the course of this study, *Z. japonica* decomposed faster than *Z. marina* in all tidal zones and bed compositions. There are several reasons why *Z. japonica* might decompose more rapidly. The simplest explanation is that, morphologically, it is smaller than *Z. marina* and thus may have a higher surface area to volume ratio. This, in turn, could lead to the higher observed rates of decomposition. An alternative explanation is that its chemical or structural composition makes it easier to break down than its native congener. Because incubation in the field was relatively short, four to six weeks, the initial degradation is the more labile compounds in the eelgrass and hence *Z. japonica* may contain more of these labile compounds while *Z. marina* contains more structural compounds. The detrital food web is important for transferring energy and nutrients from seagrasses to associated organisms (Newell 1965; Harrison and Mann 1975; Harrison 1977; Phillips 1984) and the added material contributed by *Z. japonica* is readily broken down, making it quickly available to detritivores.

On the basis of SSCU, the sediment microbial assemblage associated with *Z. japonica* differed from that of the native eelgrass. Microbial assemblages isolated from transplanted beds of invasive, native, and mixed eelgrass showed different patterns of SSCU based on the composition of the eelgrass bed. However, patterns of SSCU did not differ when compared among different tidal heights. This suggests that the vegetation is

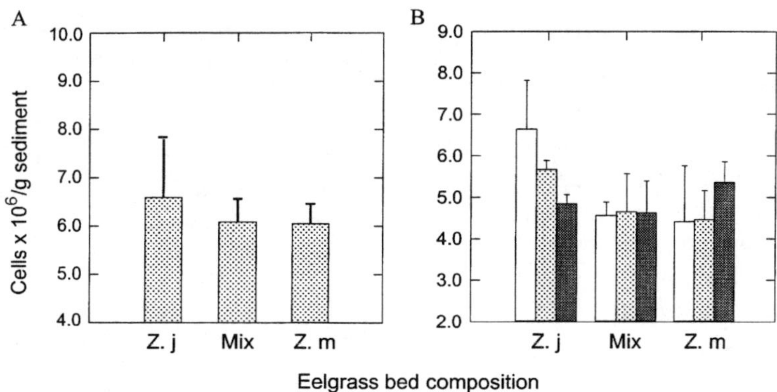

Figure 2. Abundance of bacteria from eelgrass beds of different composition as assessed by direct counts (DAPI epifluoresence microscopy). A. Naturally occurring eelgrass beds, ANOVA testing for the effect of bed composition/tidal zone: df = 2, $F = 0.21$, $P = 0.816$. B. Transplanted eelgrass beds, ANOVA testing for the effect of bed composition, tidal zone, and their interaction: bed composition – df = 2, $F = 2.92$, $P = 0.080$; tidal zone – df = 2, $F = 0.20$, $P = 0.819$; interaction – df = 4, $F = 1.25$, $P = 0.372$. Values are mean ± SE and abbreviations are as in Figure 1.

Table 2. MANOVA results (Univariate F-test) based on the first three principal components of SSCU obtained from the 31 unique carbon sources. The first three principal components explained 40% of the total variance associated with SSCU.

Effect	SS	df	MS	F	P
Test for the effect of bed composition					
Factor(1)	4.013	2	2.006	3.654	0.047
Error	9.883	18	0.549		
Factor(2)	9.994	2	4.997	5.907	0.011
Error	15.226	18	0.846		
Factor(3)	0.897	2	0.449	0.412	0.668
Error	19.589	18	1.088		
Multivariate test:					
Wilks' Lambda = 0.377		6, 32	–	3.357	0.011
Test for the effect of tidal zone					
Factor(1)	1.846	2	0.923	1.681	0.214
Error	9.883	18	0.549		
Factor(2)	0.505	2	0.253	0.299	0.745
Error	15.226	18	0.846		
Factor(3)	4.745	2	2.373	2.180	0.142
Error	19.589	18	1.088		
Multivariate test:					
Wilks' Lambda = 0.670		6, 32	–	1.180	0.342
Test for an interaction between bed composition and tidal zone					
Factor(1)	10.258	4	2.564	4.670	0.009
Error	9.883	18	0.549		
Factor(2)	0.274	4	0.069	0.081	0.987
Error	15.226	18	0.846		
Factor(3)	0.769	4	0.192	0.177	0.948
Error	19.589	18	1.088		
Multivariate test:					
Wilks' Lambda = 0.450		12, 41	–	1.251	0.282

having a strong influence on the microbial assemblage. Also of note was the interaction between bed composition and tidal height for the first principal component. The first principal component mainly comprises color development across all carbon sources and the interaction indicates that within a given bed composition, the total color development may vary among tidal zones. However, the multivariate test statistic is not significant for the interaction indicating that the overall measure of microbial community composition remains consistent within a particular bed composition across all tidal zones. Because there was no corresponding difference in microbial abundance, the differences in SSCU likely correspond to changes in microbial community composition rather than simply an alteration of microbial abundance. Changes in microbial composition could lead to different patterns or rates of nutrient mineralization, immobilization, retention, and subsequently the interaction between the microbes and the vegetation (Pedersen et al. 1999; Naeem et al. 2000).

The results of this study indicate that the abundant *Z. japonica* has the potential to alter ecosystem level processes such as decomposition and nutrient cycling in addition to the more obvious effects on habitat and infauna (e.g. Posey 1988). The rapid decomposition of the introduced eelgrass could lead to more rapid nutrient cycling which could in turn feed back into higher levels of both primary and secondary production within this system. Results also suggest that *Z. japonica* alters the associated decomposer assemblage and the functional diversity of microbes which may lead to further changes in decomposition, nutrient cycling, and nutrient retention.

Further research is needed to determine how these microbial communities differ between native and invasive eelgrass beds and to what extent the changes in the microbial assemblage affect decomposition and nutrient cycling. Additionally, the mechanism by which *Z. japonica* alters the microbial community was not elucidated in this study and warrants further investigation. In both marine and terrestrial systems, more research should focus on the effects of invaders on the microbial component of the recipient community as these microbial communities may strongly affect broader ecosystem processes.

Acknowledgements

I thank NOAA/Estuarine Reserves Division for their generous funding of this research (NERRS Fellowship Program award number NA97OR0136). I also thank Shahid Naeem for laboratory support and sound advice, Christa Fairbrother and Avalon for their valuable help in the field, Mar Wonham, Kate Howe, Gregor Schuurman and four anonymous reviewers for comments on the manuscript, the staff of the Padilla Bay National Estuarine Research Reserve for their support while on site, and those that helped to transplant vegetation.

References

Bird FL, Boon PI, and Nichols PD (2000) Physicochemical and microbial properties of burrows of the deposit-feeding thalassinidean ghost shrimp *Biffarius arenosus* (Decapoda: Callianassidae). Estuarine Coastal and Shelf Science 51: 279–291

Bulthuis DA (1991) Distribution of habitats and summer standing crop of seagrasses and macroalgae in Padilla Bay, Washington.

Technical report #2, Padilla Bay National Estuarine Research Reserve, Mount Vernon, Washington

Bulthuis DA (1995) Distribution of seagrasses in a North Puget Sound estuary: Padilla Bay, Washington, USA. Aquatic Botany 50: 99–105

Bulthuis DA (1996) Coastal habitats in Padilla Bay, Washington: a review. Technical report EL-96-15, US Army Corps of Engineers, Washington, DC

Cohen AN and Carlton JT (1995) Nonindigenous aquatic species in a United States estuary: a case study of the biological invasions of the San Francisco Bay and delta. Biological study US Fish and Wildlife Service, Washington, DC

Cohen AN and Carlton JT (1998) Accelerating invasion rate in a highly invaded estuary. Science 279: 557

Coles SL, DeFelice RC, Eldredge LG and Carlton JT (1999) Historical and recent introductions of non-indigenous marine species into Pearl Harbor, Oahu, Hawaiian Islands. Marine Biology 135: 147–158

D'Antonio CM and Vitousek PM (1992) Biological invasions by the exotic grasses, the grass/fire cycle, and global change. Annual Review of Ecology and Systematics 23: 63–87

Ellis RJ, Thompson IP and Bailey MJ (1995) Metabolic profiling as a means of characterizing plant-associated microbial communities. FEMS Microbial Ecology 16: 9–18

Elton CS (1958) The Ecology of Invasions by Animals and Plants. Methuen, London.

Garland JL (1997) Analysis and interpretation of community-level physiological profiles in microbial ecology. FEMS Microbial Ecology 24: 289–300

Garland JL and Mills AL (1991) Classification and characterization of hererotrophic microbial communities on the basis of patterns of community-level sole-carbon-source utilization. Applied and Environmental Microbiology 57: 2351–2359

Harrison PG (1976) *Zostera japonica* (Aschers. & Graebn.) in British Columbia, Canada. Syesis 9: 359–360

Harrison PG (1977) Decomposition of macrophyte detritus in seawater: effects of grazing by amphipods. Oikos 28: 165–169

Harrison PG (1982) Comparative growth of *Zostera japonica* Aschers. & Graben and *Z. marina* L. under simulated intertidal and subtidal conditions. Aquatic Botany 14: 373–379

Harrison PG and Bigley RE (1982) The recent introduction of the seagrass *Zostera japonica* Aschers. and Graebn. to the Pacific Coast of North America. Canadian Journal of Fisheries and Aquatic Science 39: 1642–1648

Harrison PG and Mann KH (1975) Detritus formation from eelgrass (*Zostera marina* L.): the relative effects of fragmentation, leaching and decay. Limnology and Oceanography 20: 924–934

Hurlbert SH (1984) Pseudoreplication and the design of ecological field experiments. Ecological Monographs 54: 187–211

Lodge DM (1993) Biological invasions: lessons for ecology. Trends in Ecology and Evolution 8: 133–137

MacIsaac HJ, Grigorovich IA, Hoyle JA, Yan ND and Panov VE (1999) Invasion of Lake Ontario by the Ponto-Caspian predatory cladoceran *Cercopagis pengoi*. Canadian Journal of Fisheries and Aquatic Sciences 56: 1

Naeem S, Hahn DR and Schuurman G (2000) Producer-decomposer co-dependency influences biodiversity effects. Nature (London) 403: 762–764

Newell R (1965) The role of detritus in the nutrition of two marine deposit feeders, the prosobranch *Hydrobia ulvae* and the bivalve *Macoma balthica*. Proceedings of the Zoological Society of London 144: 25–45

Nomme KM and Harrison PG (1991a) Evidence for interaction between the seagrasses *Zostera marina* and *Zostera japonica* on the Pacific coast of Canada. Canadian Journal of Botany 69: 2004–2010

Nomme KM and Harrison PG (1991b) A multivariate comparison of the seagrasses *Zostera marina* and *Zostera japonica* in monospecific versus mixed populations. Canadian Journal of Botany 69: 1984–1990

Oksanen L (2001) Logic of experiments in ecology: is pseudoreplication a pseudoissue? Oikos 94: 27–38

Pedersen AGU, Berntsen J and Lomstein BA (1999) The effect of eelgrass decomposition on sediment carbon and nitrogen cycling: a controlled laboratory experiment. Limnology and Oceanography 44: 1978–1992

Phillips RC (1984) The ecology of eelgrass meadows in the Pacific Northwest: a community profile. FWS/OBS-84/24, US Fish and Wildlife Service, Washington, DC

Porter K and Feig YS (1980) The use of DAPI for identifying and counting aquatic microflora. Limnology and Oceanography 25: 943–948

Posey MH (1988) Community changes associated with the spread of an introduced seagrass, *Zostera japonica*. Ecology 69: 974–983

Ruiz GM, Carlton JT, Grosholz ED and Hines AH (1997) Global invasions of marine and estuarine habitats by non-indigenous species: mechanisms, extent, and consequences. American Zoologist 37: 621–632

Ruiz GM, Fofonoff P, Hines AH and Grosholz ED (1999) Non-indigenous species as stressors in estuarine and marine communities: assessing invasion impacts and interactions. Limnology and Oceanography 44: 950–972

Stephan A, Meyer AH and Schmid B (2000) Plant diversity affects culturable soil bacteria in experimental grassland communities. Journal of Ecology 88: 988–998

Vitousek PM (1990) Biological invasions and ecosystem processes: towards an integration of population biology and ecosystem studies. Oikos 57: 7–13

Vitousek PM, Mooney HA, Lubchenco J and Melill JM (1997) Human domination of Earth's ecosystems. Science 277: 494–499

Windham L (2001) Comparison of biomass production and decomposition between Phragmites australis (common reed) and Spartina patens (salt hay grass) in brackish tidal marshes of New Jersey, USA. Wetlands 21: 179–188

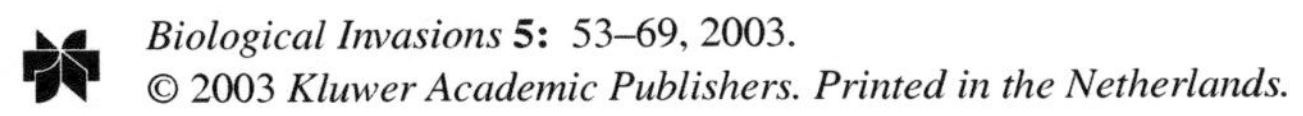
Biological Invasions 5: 53–69, 2003.

Ecological and economic implications of a tropical jellyfish invader in the Gulf of Mexico

William M. Graham[1,2,*], Daniel L. Martin[1], Darryl L. Felder[3], Vernon L. Asper[4] & Harriet M. Perry[4]
[1]*Dauphin Island Sea Lab, 101 Bienville Blvd, Dauphin Island, AL 36528, USA;* [2]*Department of Marine Sciences, University of South Alabama, Mobile, AL 36688, USA;* [3]*Department of Biology, University of Louisiana, Lafayette, LA 70504, USA;* [4]*The University of Southern Mississippi, Institute of Marine Science, Gulf Coast Research Laboratory, Ocean Springs, MS 39566, USA;* **Author for correspondence (e-mail: mgraham@disl.org; fax: +1-251-861-7540)*

Received 21 July 2001; accepted in revised form 23 January 2002

Key words: aggregations, fish eggs, predation, Rhizostomeae, scyphomedusae, zooplankton

Abstract

A large population of a previously unreported jellyfish occurred across the northern Gulf of Mexico (USA) from May through September of 2000. The jellyfish, identified as *Phyllorhiza punctata* by von Lendenfeld (1884), is not indigenous to the Gulf of Mexico or to the Atlantic Basin. Current theory states that this invasive species was introduced into the Atlantic from the Pacific Ocean through the Panama Canal about 45 years ago, and several confirmed reports indicated that a cryptic population may have existed in the northern Gulf since 1993. However, mesoscale hydrographic anomalies in spring 2000 may have directly transported the population from the Caribbean Sea where populations of *P. punctata* have been reported previously in Puerto Rico. We undertook a rapid-response sampling program from June through September to obtain ecological information regarding *P. punctata* in the highly productive northern Gulf waters. Jellyfish bell diameter increased by about 50% (from an average of 32 ± 12 to 45 ± 6 cm) as animals invaded nearshore waters during July. As the summer progressed, a net westerly distribution shift followed nearshore currents to an accumulation point near the mouth of Lake Borgne, Louisiana, in the far western Mississippi Sound. In the Lake Borgne aggregation, we estimated 5.37×10^6 medusae over an area of about 150 km^2. All of the medusae appeared to lack the symbiotic algae that are present in all other described populations, and therefore must have depended solely on planktivory for their nutrition. Clearance rates estimated from ambient zooplankton, gut contents and published digestion times ranged from $<1\,m^3\,d^{-1}$ for adult copepods to over $90\,m^3\,d^{-1}$ for fish eggs. Based on these clearance rates, the central core of this aggregation (30 km^2) was being turned over at least once per day with even higher turn-over in very concentrated 'super-swarms'. Clogging of shrimp nets was the greatest economic impact, and perhaps contributed to millions of dollars in economic losses. The indirect effect of predation on eggs and larvae of commercially important finfish and shellfish remained intangible in the determination of economic effects. In 2001, *P. punctata* has recurred along southern Louisiana and has apparently spread into the coastal and lagoonal waters of the Florida east coast.

Introduction

Introductions of gelatinous zooplankton (including ctenophores and medusae) have gained notoriety over the past two decades. Since the arrival of the North American ctenophore *Mnemiopsis leidyi* into the Black Sea in the early 1980s (Kideys 1994; Shiganova 1998; Shiganova and Bulgakova 2000) and more recently into the Caspian Sea (Ivanov et al. 2000), research has routinely pointed to the controls that gelatinous zooplankton can exert on marine planktonic foodwebs, including important commercial fisheries (Purcell 1985, 1989; Purcell and Arai 2001; Purcell et al. 2001). The possibility that exotic medusae can become established in regions of important commercial fisheries is of particular concern because feeding rates, especially on eggs and larvae of commercially important fish are typically high (deLafontaine and Leggett 1988; Cowan and Houde 1992; Arai 1997; Purcell 1997; Purcell and Arai 2001).

A large population of a previously unreported jellyfish occurred across the northern Gulf of Mexico (hereafter referred to as the northern Gulf) from May through September of 2000. As summer progressed, scientists, resource managers, commercial fishers, residents, and tourists along the northern Gulf realized that the extent and magnitude of this new jellyfish was extraordinary. The juxtaposition of massive numbers of these unusually large, spotted jellyfish with fishing activity and fisheries resources created concern that negative economic as well as ecological impacts were developing. The jellyfish, identified as *Phyllorhiza punctata* von Lendenfeld (1884), a scyphomedusa of the Order Rhizostomeae and Family Mastigiidae, was not indigenous to the Gulf of Mexico, or even to the Atlantic Basin. Current theory suggests that *P. punctata* was introduced into the Atlantic Basin from the Pacific Ocean through ship-mediated transport at least 45 years ago (da Silveira and Cornelius 2000). Initial historical reports from the Caribbean in 1955 by Moreira (1961) gave these medusae the pseudonym of *Mastigias scintillae.*

Global spread of a tropical invader

P. punctata, indigenous to the tropical western Pacific Ocean, has been a successful migrant around the world (Figures 1a, b). Though von Lendenfeld (1884) originally described Port Jackson, Australia as the 'type' locality, it has undoubtedly occurred widely in Australian coastal and lagoonal waters and throughout the Indo-Pacific Ocean including the Philippine archipelago (Heeger et al. 1992) and Thailand (P. Cornelius personal communication). Beyond its 'native' distribution of the Indo-Pacific, *P. punctata* has been successful at invading the central (Devaney and Eldredge 1977; Clarke and Abey 1998) and eastern Pacific (Larson and Arneson 1990), the Mediterranean Sea (Galil et al. 1990), the southwestern Atlantic Ocean (Mianzan and Cornelius 1999) and the northern (Cutress 1971; Garcia 1990; Garcia and Durbin 1993) and southern Caribbean Sea (Moreira 1961; da Silveira and Cornelius 2000).

We present here the first detailed information on the extent and magnitude of *P. punctata* in the Gulf of Mexico. Though this bloom of jellyfish was entirely unforeseen, we hastily organized and implemented a sampling program, often taking advantage of fortuitous field opportunities, to collect as much information as possible to describe characteristics of feeding, growth, and reproductive potential. These data are discussed by comparing northern Gulf *P. punctata* to other populations and by comparing potential impacts of *P. punctata* to native Gulf of Mexico jellyfish. In addition, we present a brief summary of the realized and anticipated economic impacts that were (or might be) experienced should *P. punctata* persist in the northern Gulf.

Identification

Accurate identification is a major caveat in discussing the global spread of any invasive species, and no taxonomic group is more troubling than the Phylum Cnidaria. A great deal of confusion has surrounded the Mastigiidae (see Cutress 1971; Larson and Arneson 1990; da Silveira and Cornelius 2000). Until a comprehensive taxonomic revision of the Rhizostomeae is available, we made our identification based on two lines of evidence. Firstly, with the exception of size and pigmentation (i.e. apparent lack of zooxanthellae in the northern Gulf population; Figures 2a, b), we confirmed all other characteristics considered diagnostic of *P. punctata* in Mayer's (1910) original monograph. Later synopses by Kramp (1961), and more recently compiled by Mianzan and Cornelius (1999), were also consistent with the generic placement. Specific diagnosis of the northern Gulf population include: (1) 8 rhopalia; (2) 14 lappets in each octant arranged as 4 sets of double lappets in the middle of each octant, flanked by single lappets; (3) sub-genital ostia wider than high; (4) circular sub-umbrellar muscles interrupted by the 8 radial canals; (5) 8 radial canals communicate directly with the stomach; (6) white crystalline inclusions creating spots or 'punctae' on the surface of a finely granular bell (Figure 2b). Kramp (1961) and da Silveira and Cornelius (2000) differentiate *Phyllorhiza* spp. from *Mastigias* spp. by the large 'windows' found in the lateral membranes of the mouth-arms of the

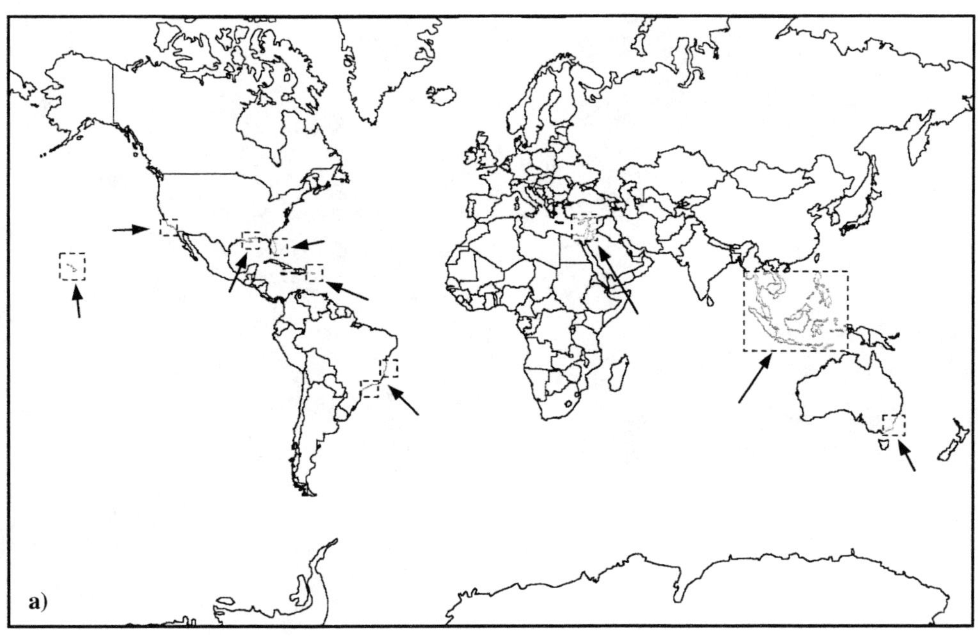

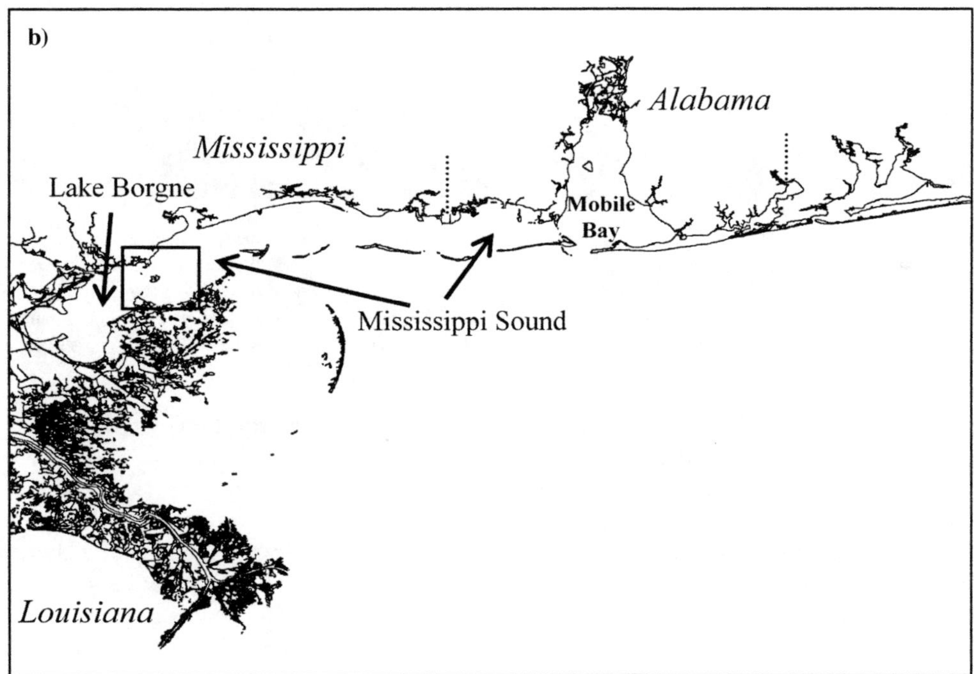

Figure 1. Maps of pertinent study regions showing (a) type localities and translocation history of *P. punctata*, and (b) region of this study in the northern Gulf of Mexico including Mississippi Sound as indicated by the arrows. The region enclosed by the box indicates the location of the jellyfish aggregation referred to in Figure 5.

former. In addition, sperm ultra-structure in Gulf of Mexico *P. punctata* (W. Graham and A. Moss, unpub. data) is identical with that described by Rouse and Pitt (2000) for Australian *P. punctata*. A second line of evidence used here is the historical account of *P. punctata* introduction in the Atlantic Ocean (Figure 1a). *P. punctata* is the only species of Mastigiidae documented from the western Atlantic Ocean

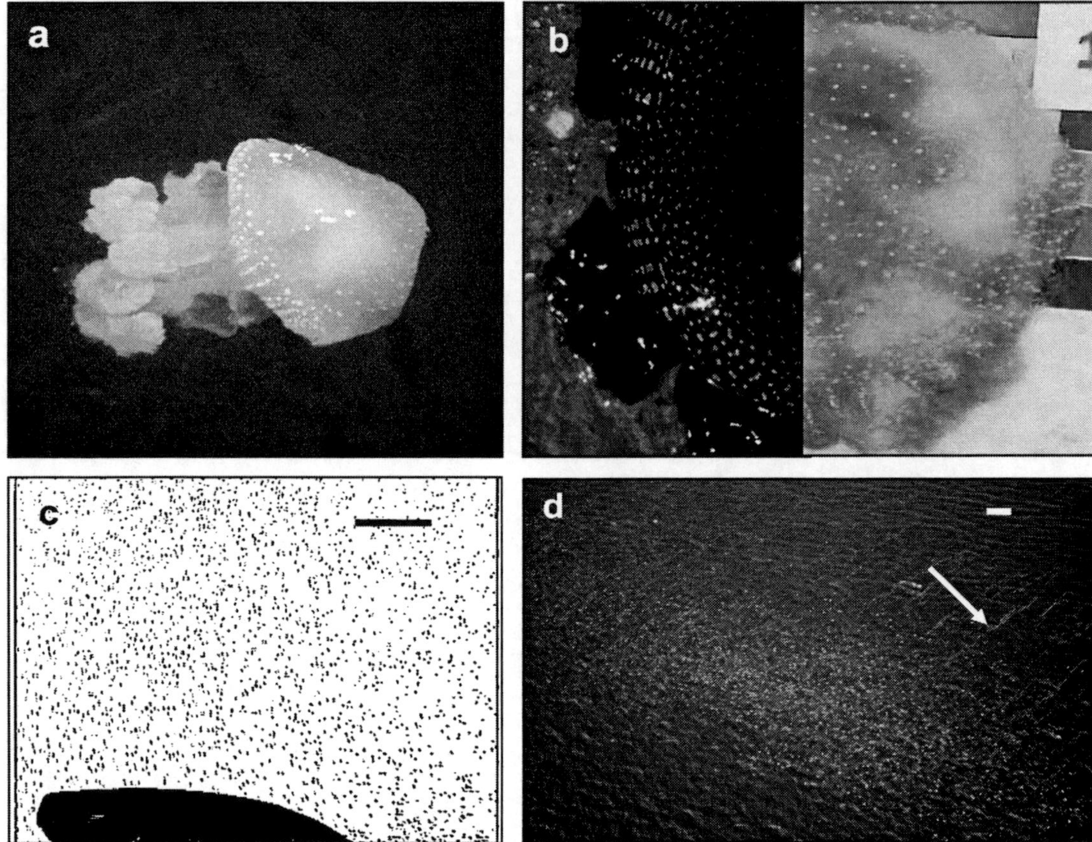

Figure 2. Images of *P. punctata* individuals and aggregations during summer 2000. (a) A medusa approximately 50 cm in bell diameter swimming at the sea surface; (b) A composite image showing zooxanthellate *P. punctata* from Western Australia on the left (photo courtesy of Dr R. Rippingale) with a bleached medusa from the northern Gulf in 2000; (c) aerial image of medusae in the central core of the Lake Borgne aggregation with background water masked to reveal only the medusae as black spots (the large black area is the airplane's wingtip), and (d) a super-swarm of medusae from Lake Borgne, Louisiana, with foam streaks (arrow) emanating from the swarm. The scale bars in c–d are 20 m in length.

(see Kramp 1961; Mianzan and Cornelius 1999), and as such it is reasonable to presume that hydrographic connectivity within the tropical Atlantic Ocean and the Caribbean Basin has, in part, allowed for expansion of the previously reported populations of this species to spread both northward and southward (Figure 1a).

The identification we have made must, however, be taken with a degree of caution as morphology-based systematics within the Scyphozoa are outdated, subject to misinterpretation and in dire need of revision. It is likely that only extensive molecular analyses will ultimately confirm this species. While the identification of this species may change in the future, the Mastigiidae – of which this is certainly a member – radiated in the tropical western Pacific, and as such the invasiveness of this species will remain valid.

Materials and methods

Distribution surveys

All surveys were conducted between 3 July and 28 September 2000. Anecdotal reports prior to (and occasionally within) this time frame were also considered only if the sources were associated with marine research or resource management. Three types of surveys were conducted. First, a total of 13 small-boat sampling trips were conducted between 3 July and 27 July. Nine of these were in coastal Alabama extending offshore no more than 20 km. Four sampling trips were in Mississippi coastal waters including Mississippi Sound and its bordering barrier islands (Figure 1b). During all of these local surveys, we focused on the distribution of medusa patches along the mainland coastline, barrier islands, and inlets. Second, a series of six aerial surveys were conducted over most of the entire coastal region between Mobile Bay, Alabama and the far western Mississippi Sound. Two of these flights (in early July and mid-August) were conducted by the Mississippi Department of Natural Resources as part of ongoing fisheries activity surveys; four flights were conducted specifically for this project. All flights were made in a single-engine Cessna – a small plane with overhead wings – that allowed us to make observations and collect imagery directly below the plane. Aerial surveys were appropriate here because the shallow water depths as *P. punctata* swimming behavior (i.e. milling at the surface and shallow dives; W.M. Graham and D.L. Martin, unpub. obs.) tended to keep the animals visible from the air. Third, a comprehensive large vessel survey was conducted on 11–12 August using the *R/V Pelican* (Louisiana Universities Marine Consortium) in the vicinity of Lake Borgne, Louisiana (Figure 1b).

Estimating jellyfish density

On 2 August and 11 August, aerial images of the largest aggregation at the entrance to Lake Borgne, Louisiana, were collected in order to estimate surface densities of *P. punctata*. On 2 August, 35 mm still photographic images, taken in slide format, were subsequently digitized for image processing. On 11 August, digital video sequences of various surface concentrations of jellyfish were captured with a digital video camera. All imagery was taken out of the passenger side window during steep banks in order to provide a vertical aspect. All passes were made over jellyfish aggregations at altitudes between 100 and 325 m above sea level depending on format and resolution. Spatial calibration was made by either the crude method of capturing the aircraft's shadow over water or by capturing known, measurable objects on land (e.g. ground-truthed stripe spacing of a parking lot). Care was taken to make separate calibration passes at the same altitude and camera magnification as the jellyfish passes.

All image processing was done at Dauphin Island Sea Lab (DISL). Still images were digitally scanned and converted to RGB format for further processing. The digital images were processed using ImagePro Plus imaging software. Two of the images used in this analysis are presented in Figures 2c, d. These images were processed by masking the background water (see Figure 2c) and then counting medusae in the final masked and calibrated images. When possible, discrete jellyfish were counted. However, images of the most concentrated aggregations often did not produce discrete data because the medusae were overlapped in the images (e.g. Figure 2d). For these images, we calculated percent coverage of jellyfish pixels in the images and applied a conversion of pixels jellyfish^{-1} determined from the discrete jellyfish.

The *R/V Pelican* cruise of 11–12 August coincided with an imaging survey described above. Overall aggregation size and density was estimated by combining the ship-based visual survey counts with the calibrated aerial imagery to produce a map of jellyfish distribution and concentration. The visual survey was conducted by counting all individuals in 5 min intervals passing within approximately 50 m of either side of the ship. This survey was conducted continuously for about 5 h as the ship steamed in and out of the aggregation. Counts were subsequently scored as 'absent', 'low', 'moderate', or 'high' concentrations. Actual densities within each concentration grouping were measured from the calibrated aerial surveys, and were fitted into this surface map to estimate overall numbers within the aggregation.

Field collections and processing

On 10 dates between 3 July and 12 August, a total of 96 medusae were collected for size-distribution along coastal Alabama and Mississippi. These medusae were measured to the nearest 1 cm across the flattened bell by placing the animal aboral-side down. On 3 July and 7 July, sex was determined for 20 haphazardly selected medusae by taking a biopsy of gonad material and viewing it by microscopy in the laboratory. Medusa diameter was converted to wet mass using our own empirically derived formula from 35 medusae of varying size:

$$M_w = 9.89 * D^{1.731}, \qquad (1)$$

where M_w is the wet mass in g and D is diameter in cm. Further conversions for dry weight, percent water content, percent ash and ash-free dry weight (AFDW) were derived from 25 medusae by sectioning approximately 10 g of bell and drying at 60 °C to constant weight (at least 48 h). Dry weight and water content was determined by difference after correcting for a 10% 'water of hydration' (Larson 1986). Percent ash and AFDW was determined by combusting the dried tissue overnight at 500 °C and then re-weighing the remaining ash. Because water salinity influences salt (and ash) content, and thus percent organic content, we collected salinity information on these three dates using a SeaBird SBE25 conductivity–temperature–depth probe (CTD). On 11–12 August, 17 medusae ranging 34–52 cm were processed for carbon and nitrogen content. Bell margin tissue samples were frozen in liquid nitrogen at sea, lyophilized at Dauphin Island Sea Lab, and then measured for both C and N using a Carlo-Erba NA1500 CNS analyzer.

Throughout the summer of 2000, we assessed the status of zooxanthellae in freshly caught jellyfish. Initially we excised epithelial tissue from the umbrella, mouth-arms and sub-umbrellar muscle bands and examined each microscopically for presence of zooxanthellae. Since no zooxanthellae or coloration were noted initially, we subsequently used only macroscopic evidence of any color variation through the remainder of 2000 (see Figure 2b, right panel).

Gut content analysis was performed on 11–12 August for 17 medusae ranging 34–52 cm collected at three separate sites within the Lake Borgne, Louisiana, aggregation. We followed the basic procedure of Graham and Kroutil (2001) and excised the guts to collect undigested and partially digested material on a 200-μm mesh sieve. A smaller mesh more suitable for retention of microplankton (<200 μm) could not be used due to clogging by copious amounts of mucus produced by these medusae. Sieved samples were concentrated and fixed in 5% borate-buffered formalin in filtered seawater, and identified and enumerated at Dauphin Island Sea Lab.

Zooplankton tows

Duplicate zooplankton tows were made at six stations (3 inside and 3 outside the jellyfish aggregation) using a 153-μm mesh, 75 cm diameter ring net towed obliquely over the whole water column (approximately 3 m vertical depth) for about 1 min. The net was fitted with a calibrated flow meter. One-half of this material was immediately fractioned into >1000 μm and 153–1000 μm size-classes. Two quantitative splits (at least 1/32 of the total sample) were dried for 24 h on pre-weighed Nitex filters to determine dry mass. The other one-half split was immediately fixed in 5% borate-buffered formalin in seawater and returned to Dauphin Island Sea Lab where they were identified and enumerated.

Results

Distribution and movement of P. punctata *in the northern Gulf of Mexico*

Second-hand reports of large, 'unusual' jellyfish (presumably *P. punctata*) were first made during the last week of May and continued into mid-June. Two of these reports placed highest concentrations of animals along strong frontal features about 20–40 km south of Mobile Bay, Alabama (Figure 3a). In addition, these reports usually stated that the fronts were also accumulating large rafts of sargassum weed. In mid-June as *P. punctata* populations were moving into Mobile Bay and Mississippi Sound, rafts of sargassum were washing onto local beaches.

The first systematic sampling in mid-July found *P. punctata* distributed widely along the barrier islands of southern Mississippi Sound (Figure 3). Highest concentrations were in the inlets between the islands and along western points of several islands. During this period, Louisiana fisheries resource scientists also indicated 'large numbers' of *P. punctata* near Port Fourchon in south-central Louisiana (Figure 3b). By mid-July we observed concentrations of jellyfish along north and south shores of the barrier islands, and northward into Mobile Bay (Figure 3c). By mid-August a net westward drift of the population toward an accumulating point near Lake Borgne, Louisiana, had occurred (Figures 3c–e). Densities always remained highest near the entrance to Lake Borgne, Louisiana, and concentrations gradually declined at the eastern extent in Mobile Bay. By early to mid-September, only sparse individuals, and not aggregations, occurred along the barrier island chain (Figure 3f). A modest aggregation of medusae remained in the far western Mississippi Sound and along the northern Chandeleur Islands in mid-September. By the end of September, only a few individuals remained in the westernmost Mississippi Sound (Figure 3g). Interestingly there was a single anecdotal report of large numbers of dead *P. punctata* washed-up in the northern reaches of the Mobile Bay estuary (Figure 3g). No observations or confirmed reports of *P. punctata* were made in the northern Gulf of Mexico after late September. In summary, aggregations in Mississippi Sound were never found along northern sound shores, but were most concentrated within the barrier island passes and within a 10-km band extending north and south of the islands. No medusae were ever found or reported east of Mobile Bay, Alabama in 2000.

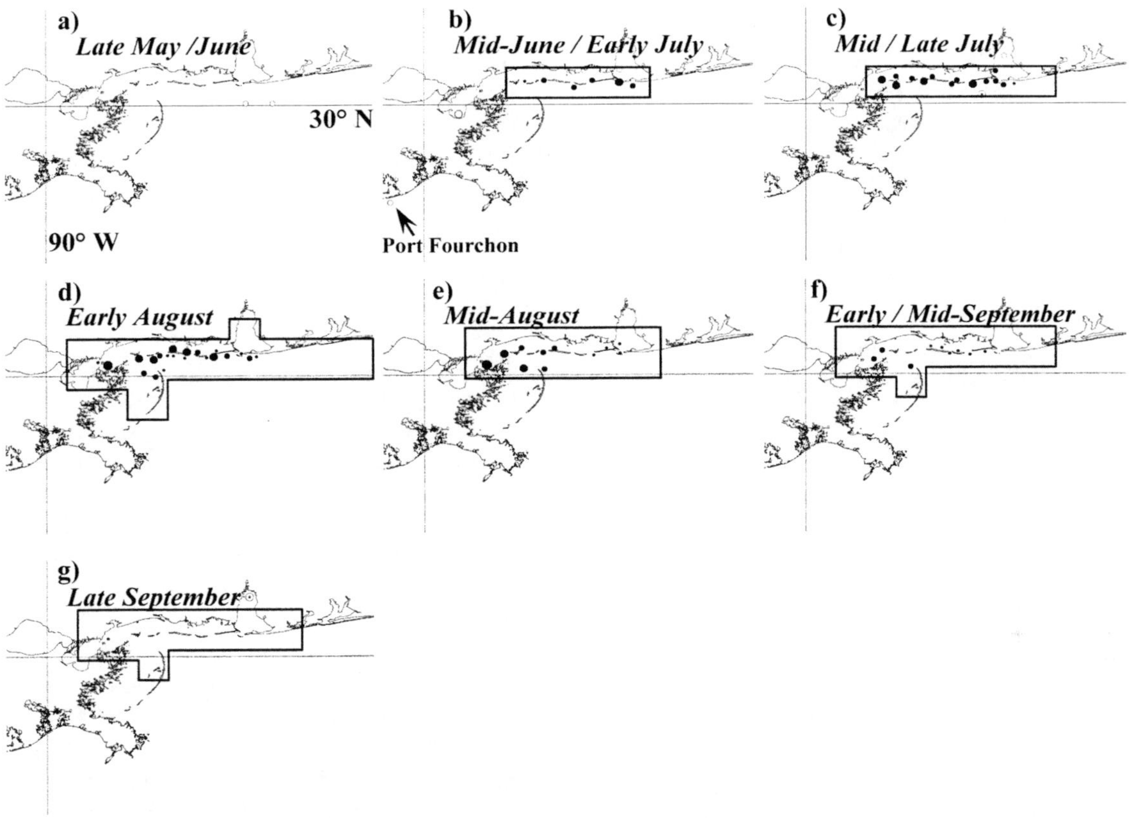

a)
Late May /June
30° N
90° W
b)
Mid-June / Early July
Port Fourchon
c)
Mid / Late July
d)
Early August
e)
Mid-August
f)
Early / Mid-September
g)
Late September

Figure 3. Composite distribution map of *P. punctata* in the northern Gulf of Mexico in 2000. The boxes in each panel indicate the survey extent during that time frame. Size of symbol indicates relative density of medusae, and open symbols are from anecdotal information, while filled symbols are our own observations. Open symbol with a star indicates anecdotal information of only dead medusae.

Size distribution

The average bell diameter of *P. punctata* increased by about 50% during July (Figure 4). Prior to July 15, average bell diameter was 32 ± 12 cm. With the exception of one sampling trip when average size was 27 cm, all other daily averaged bell diameters were about 35 cm. After July 15, mean bell diameter increased to 45 ± 6 cm, significantly larger than the pre-July 15 mean size (t-test, $t = -5.54$, $P \ll 0.001$). Between 18 July and 12 August, average bell diameter consistently remained around 45 cm with little variation between sampling dates.

Bell Diameter (cm)
60
55
50
45
40
35
30
25
20
6/28
7/3
7/8
7/13
7/18
7/23
7/28
8/2
8/7
8/12
8/17
Sampling Date

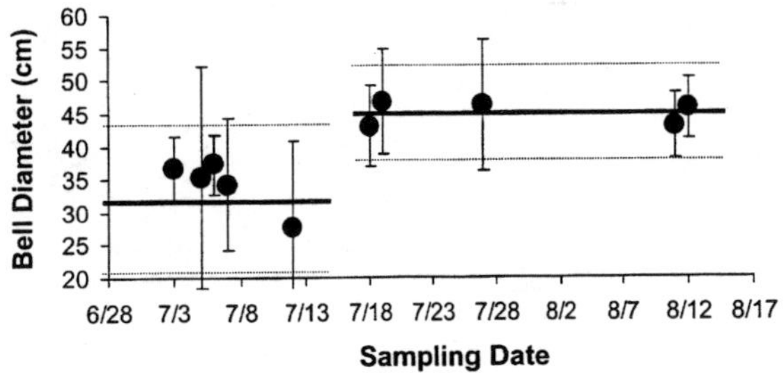

Figure 4. *P. punctata* bell size distribution as measured by bell diameter. Means and standard deviations are presented for each sampling date (circles), and an overall mean and standard deviation before and after 15 July is indicated by horizontal lines.

Jellyfish composition

A one-way ANOVA revealed that water content did not vary significantly between different ambient salinities when or where medusae were collected (ranging 29.0–32.5‰ over 5 separate sampling dates). Mean water content of *P. punctata* was 96.2 ± 0.25% (Table 1). This included a correction (10% of dry mass) for 'bound' water that Larson (1986) termed 'water of hydration'. Like water content, percent ash content of corrected dry weight did not significantly vary over the salinity regimes of time or place of collection.

Inorganic ash constituted on average 77.1 ± 4.5% of tissue dry mass. Combustible carbon and nitrogen was 12.0 ± 6.9% and 3.0 ± 1.5% of the dry mass, respectively (Table 1). *P. punctata* umbrellar tissues, had an overall C : N ratio of 4 : 1.

Initial microscopy revealed that zooxanthellae were absent from all tissue types. At no other point did coloration qualitatively vary from the initial population, and we assumed that the entire northern Gulf population lacked zooxanthellae throughout 2000.

One of the most interesting initial findings was that 100% of the medusae analyzed for sex ($n = 20$) were male. All medusae assessed had large quantities of live sperm that emanated from numerous follicles in the gonadal tissue. Using wet-mount microscopic techniques, we observed that sperm released from the follicles quickly aggregated into spermatozeugmata (see Arai 1997). No ova or developing planulae were recovered from medusae.

Table 1. P. punctata bell tissue composition and conversion factors, mean and SD.

Measured property	Mean	SD
% Water	0.962	0.0025
Ash : dry mass	0.771	0.045
C : dry mass	0.120	0.069
N : dry mass	0.030	0.015
Overall C : N	4.0	—

Percent water and percent ash of dry mass was determined from a total of 25 medusae of varying sizes collected in Alabama and Mississippi coastal waters. Carbon and nitrogen content were measured from 17 medusae of varying sizes collected in Lake Borgne, Louisiana.

Quantification of the Lake Borgne aggregation

The Lake Borgne aggregation *of P. punctata*, which probably formed by early July according to anecdotal information from Louisiana marine resources managers (Figure 3b), was the subject of intense study over a two day period 11–12 August 2000. Synoptic ship and airplane surveys revealed a large ovoid aggregation approximately 150 km^2 in size (Figure 5). Ship-based visual scoring into 'low', 'medium' and 'high' concentrations identified an outer, low density halo of approximately 120 km^2 surrounding an inner core ('medium' and 'high' concentrations) of approximately 30 km^2. Within this central core, an area of approximately 1 km^2 was occupied by highly concentrated patches of medusae (Figure 5). To distinguish these inner patches from the overall aggregation and from the central core, we referred to them as 'super-swarms'. Visualized from

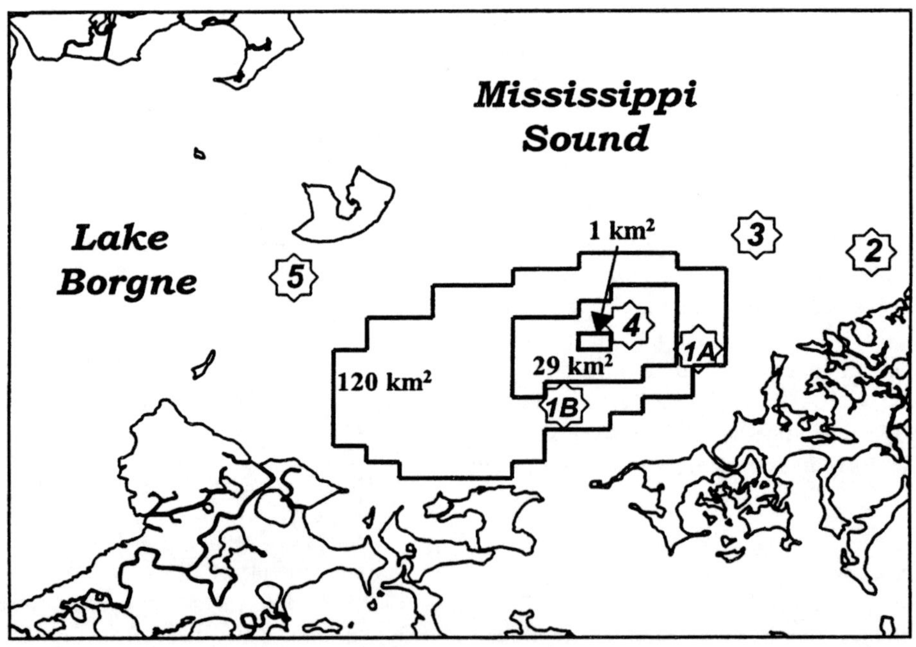

Figure 5. Composite distribution map of the Lake Borgne aggregation. The estimated total areas of 'low', 'medium' and 'high' jellyfish densities are indicated. The locations of 6 sampling stations are also indicated relative to jellyfish densities where they were occupied.

the air, super-swarms were on the order of 1000 m long and 100–200 m wide (see Figure 2d). Several of these patches typically repeated over 1–2 km of the central core. Outside of the super-swarms, densities scored as 'medium' in the visual surveys (29 km^2 total area) contained a qualitatively uniformly dispersed aggregation of medusae (Figure 2c).

Areal jellyfish densities estimated using image analysis were $0.0024\,m^{-2}$ in the low density halo, $0.094\,m^{-2}$ in the central core, and $2.35\,m^{-2}$ in the highest density super-swarms (Table 2). We applied these measured jellyfish concentrations to the total area estimated from the visual survey to estimate 2.88×10^5 medusae in the outer halo, and 5.01×10^6 in the central core. Within the central core, super-swarms accounted for an estimated 2.35×10^6 medusae (Table 2). By applying the empirically-derived conversions reported in Table 1, we estimated the Lake Borgne aggregation, with an average medusa size of 44.6 cm, had a total wet biomass of 37.0×10^6 kg, and a dry mass of 1.44×10^6 kg. In addition, the aggregation represented a 1.73×10^5 kg pool of carbon and a 4.32×10^4 kg pool of nitrogen (Table 2).

Gut contents

Nearly all (96%) of the gut contents from the 15 processed jellyfish were represented by only four taxonomic groups (Figure 6). Over the three stations (1A, 1B, and 4 in Figure 5), ranked numerical dominance of prey taxa was bivalve larvae (35%), adult copepods (23%), loricate tintinnids (23%), and fish eggs (15%). Tintinnids (approximately 150 μm long and 50 μm wide) were likely underestimated because of the relatively large mesh size (200 μm) used to retain the gut contents during processing. On average, *P. punctata* sampled at station 4 had an order of magnitude more prey items in their guts (1651 prey) than at station 1A or 1B (294 and 128 prey, respectively). This pattern held across all prey taxa with the exception of adult copepods at station 1A (Figure 6).

Ambient zooplankton abundance

While total zooplankton dry mass was not depressed within the jellyfish aggregation (stations 1A, 1B and 4 in Figure 5), the <1 mm size-fraction (shaded bars in Figure 7a) was lower within the aggregation when compared with the two stations outside of the aggregation by a factor of about 5. As seen in Figure 7b, zooplankton concentrations were consistently depressed inside the aggregation. Also as indicated in Figure 7b, the lowest concentration of mesozooplankton was at station 4, which had the highest measured concentration of *P. punctata*. Adult copepods were the dominant mesozooplankton (represented primarily by the calanoid *Acartia tonsa*). As seen in Figure 7b, variations in copepod abundances tracked the overall pattern of mesozooplankton among stations.

Estimated clearance potential

First-order estimates of taxon-specific prey clearance rates (F) were derived using the following equation:

$$F = H/A \times D, \quad (2)$$

where H is the concentration of a specific prey taxon in the jellyfish gut, A is the ambient concentration of the same prey taxon, and D is the prey digestion time. Estimates were made for adult copepods, fish eggs and

Table 2. Measured and calculated properties of the Lake Borgne, Louisiana, *P. punctata* aggregation. The observation concentration scores were made over 11–12 August 2000 from the research ship.

Concentration score	Area (km^2)	Concentration (m^{-2})	#Medusae ($\times 10^6$)	Aggregation biomass ($\times 10^6$ kg) Wet	Dry	Ash-free	C	N	Water column turnover (d^{-1})
Low	120	0.0024	0.288	2.04	0.0775	0.0177	0.00930	0.00232	0.0062–0.074
Medium	29	0.094	2.73	19.3	0.733	0.168	0.0880	0.0220	0.24–2.9
High	1	2.35	2.35	16.6	0.631	0.144	0.0757	0.0189	6.0–72.5
Cumulative	150		5.37	37.9	1.44	0.330	0.173	0.0432	

These estimates were combined with measured concentrations collected by aerial survey to yield total number of medusae in each region and in the whole aggregation. Conversions to dry mass, carbon and nitrogen were made using values from Table 1, and the dry mass includes a 10% correction for water of hydration (Larson 1986). Estimated turn-over was based on volumetric concentration assuming a uniform water column depth of 3 m in Lake Borgne and an average medusa size of 44.6 cm (=7.082 kg wet mass). The range of values reflects low and high clearance rate estimates presented in Table 3.

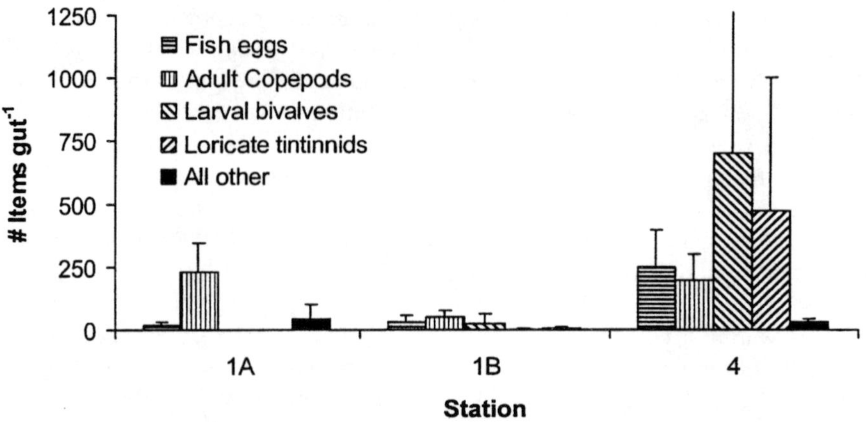

Figure 6. Average number of prey in guts of *P. punctata* from the Lake Borgne aggregation. Three stations were sampled, error bars indicate standard deviation.

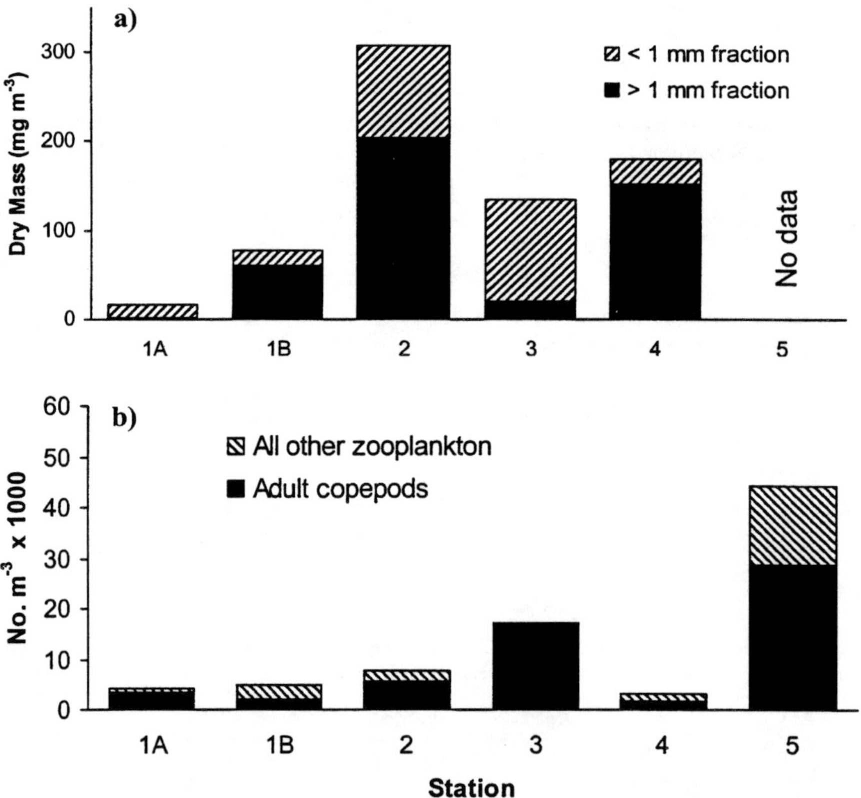

Figure 7. Ambient zooplankton concentrations (>200 μm) in the region of Lake Borgne, Louisiana. Refer to Figure 5 for station locations. (a) Size-fractionated zooplankton dry mass, no dry mass data were collected at station 5. (b) Concentration of total zooplankton, and the adult copepod fraction of the total zooplankton assemblage.

larval bivalves at the three Lake Borgne stations with *P. punctata* (Table 3). Since digestion times were not measured, we used published digestion times reported for other rhizostome medusae. Copepods were digested by *Pseudorhiza haekeli* in 2 h (Fancett 1988), fish eggs were digested by *P. haekeli* in 3 h (Fancett 1988), and bivalve larvae were digested by *S. meleagris* in 4 h (Larson 1991). Estimated clearance rates were lowest for copepods (0.8–1.2 $m^3 d^{-1}$, mean = 0.77 $m^3 d^{-1}$), followed by larval bivalves (0.0–22.8 $m^3 d^{-1}$, mean = 7.7 $m^3 d^{-1}$), and fish egg clearance rates were the greatest (3.8–229.6 $m^3 d^{-1}$, mean = 92.5 $m^3 d^{-1}$) (Table 3).

Discussion

Ecological considerations

Consumption by *P. punctata* on zooplankton in the northern Gulf of Mexico was limited to a narrow window of prey types (Figure 6). Gut contents of medusae from Lake Borgne showed surprisingly low diversity relative to the overall ambient mesozooplankton community typically found in coastal regions. Of 19 taxa enumerated from the ambient water column, only 3 of these taxa and tintinnids, which were too small to be enumerated from the >153 μm net samples, constituted >95% of the gut contents. Similarly, Larson (1991) reported that >90% of consumed prey by native Gulf of Mexico rhizostome medusae *S. meleagris* belonged to only 6 taxa including, in order of abundance, bivalve veligers, copepod nauplii, other gastropod veligers, calanoid copepods, and tintinnids. While there is almost complete overlap in diet between native and non-native rhizostome medusae, we found relatively few copepod nauplii, which could have been either an artifact of sampling or, more likely, size-based differences in relative abilities to capture and retain different zooplankton prey taxa (Suchman and Sullivan 1998; Graham and Kroutil 2001).

Among those dominant taxa recovered from the guts, it appears that consumption of the smaller, slower swimming microzooplankton and non-motile fish eggs was key to the daily energetic requirements of *P. punctata* in this region. The large numbers of tintinnids recovered from *P. punctata* at Station 4 in Lake Borgne (Figure 6) not only reflected a sizable proportion of the gut contents, but also a high quality source of nitrogen with C : N values around 4 : 0 (Stoecker and Govoni 1984). Microzooplankton grazing by *P. punctata* likely reflects the evolutionary history of the Rhizostomeae, which radiated in the tropics (Kramp 1970) where oligotrophy yields microbial-dominated trophic pathways. Therefore, the expansion of *P. punctata* into the very productive northern Gulf of Mexico coastal waters is exceptionally interesting because this invasive species can apparently bridge productivity regimes.

While we did not specifically target the microzooplankton size-class as part of the initial response to *P. punctata* in 2000, cell densities of microzooplankton in the eutrophic northern Gulf of Mexico have been reported as 20–30 × 10^6 cells m^{-3} (Strom and Strom 1996). In southern Louisiana, Dagg (1995) estimated that as much as 95% of phytoplankton grazing is by the microzooplankton community. Owing to the high growth rates of microzooplankton, it is unlikely that substantial turnover due to *P. punctata* had occurred in 2000. However, the reliance on microzooplankton by *P. punctata* was likely an important transfer of energy into the largely remineralizing jellyfish component of the food web.

It is interesting to note that both the native *S. meleagris* (Larson 1991) and the invasive *P. punctata* were inefficient consumers of adult copepods (Figure 6, Table 3). Costello and Colin (1994, 1995) explained that differential ingestion of prey is a function of overlap between marginal bell velocities of the feeding jellyfish and the swimming velocities of the prey.

Table 3. Estimation of clearance potential of adult copepods, fish eggs and larval bivalves by *P. punctata* from three stations in the Lake Borgne aggregation, 11–12 August 2000.

Station	Adult copes[a]			Fish eggs[b]			Larval bivalves[c]		
	Gut^{-1}	Ambient (m^{-3})	Clearance ($m^3 d^{-1}$)	Gut^{-1}	Ambient (m^{-3})	Clearance ($m^3 d^{-1}$)	Gut^{-1}	Ambient (m^{-3})	Clearance ($m^3 d^{-1}$)
1A	230.9	3346.0	0.8	17.2	36.5	3.8	0.0	8.3	0.0
1B	53.9	2169.1	0.3	34.4	1.2	229.6	28.5	334.7	0.34
4	197.3	1918.9	1.2	251.3	45.5	44.0	700.0	123.8	22.8
Average			0.77			92.5			7.7

Refer to Figure 5 for station locations. Published digestion times are indicated for each prey type.
[a]Copepods digestion: 2 h (Fancett 1988, *P. haekeli*).
[b]Fish eggs digestion: 3 h (Fancett 1988, *Pseudorhiza haekeli*).
[c]Bivalve larvae digesion: 4 h (Larson 1991, *Stomolophus meleagris*).

Rhizostomes require entrainment of prey on a fine mesh-like network of feeding 'mouthlets' located along the 8 branching mouth-arms, which is in contrast to the entanglement of prey in the long tentacles of semaeostome medusae. Capture of prey by both *S. meleagris* and *P. punctata* appeared to be limited to smaller-sized (e.g. microzooplankton) and non-motile (e.g. fish eggs) zooplankton. Recent laboratory experimentation using 1.5–7.5 cm *P. punctata* by D'Ambra et al. (2001) showed that adult copepod swimming velocities were adequate to escape capture by the medusae. In contrast, Garcia and Durbin (1993) reported that the copepod *Acartia tonsa* was the dominant prey of *P. punctata* in Laguna Joyuda, Puerto Rico. Although clearance rate of copepods was lower than fish eggs and larval bivalves, adult copepod biomass was still reduced in the Lake Borgne aggregation suggesting at least a modest impact on copepods by *P. punctata* (Figure 7b).

In all other described populations, from both native and non-native locales, *P. punctata* is host to highly concentrated algal symbionts. In Australia, these zooxanthellae account for its common name, the 'brown jellyfish' (see Figure 2b). The presence of zooxanthellae is widespread among tropical rhizostome medusae, particularly among the Mastigiidae, presumably as a means to compensate for low secondary production of zooplankton prey (Arai 1997). While zooplankton consumption by *P. punctata* in Laguna Joyuda, Puerto Rico, was considered a supplemental energy source to algal photosynthate (Garcia and Durbin 1993), lack of zooxanthellae in the northern Gulf of Mexico population is evidence of diet-induced morphological plasticity in *P. punctata*. The lack of zooxanthellae in the northern Gulf of Mexico also created a practical problem in the initial identification of the species since zooxanthellae and coloration had been previously used as diagnostic characteristics in *P. punctata* (Mianzan and Cornelius 1999).

Lack of zooxanthellae was certainly not inhibitory to success of the species in turbid, light-limited northern Gulf waters. Loss of zooxanthellae as an energy source was particularly intriguing since energetic requirements must have been met completely by zooplanktivory. Despite the presumed disadvantage of losing algal symbionts, northern Gulf *P. punctata* were considerably larger than other described populations. Without zooxanthellae, *P. punctata* in the Gulf of Mexico obtained an average bell diameter of 45 cm (maximum 62 cm) (Figure 4). The same species in the Caribbean and Australia with zooxanthellae were substantially smaller. Individuals in the Caribbean were reported to be 24 cm diameter by Garcia and Durbin (1993) and 33 cm by da Silveira and Cornelius (2000) (a 50 cm maximum size described by Cutress (1971) was not supported by data). Likewise, maximum sizes were considerably smaller in the Pacific: about 30 cm in western Australia (Rippingale and Kelly 1995), 40 cm in Hawaii (Clarke and Abey 1998), and 7 cm (one individual) in southern California (Larson and Arneson 1990). Because growth, size and maturation are food-limited (e.g. Olesen et al. 1994; Ishii and Bamstedt 1998; Lucas and Lawes 1998), the large size of the Gulf of Mexico individuals suggests that energetic needs are being met or exceeded by grazing on the regionally very high rates of secondary production of the northern Gulf. Garcia and Durbin (1993) reported a daily ration of 3.1–6.4% of the AFDW of *P. punctata* in Laguna Joyuda, Puerto Rico. However, the Laguna Joyuda population contained symbionts which provided most of the daily energetic requirements (Garcia and Durbin 1993).

In addition to direct effects of predation on copepods, we believe *P. punctata* may have had an indirect effect on zooplankton production through changes in chemical or physical properties of the water. The manifestation of surface foam streaks down-wind of a super-swarm (see arrow in Figure 2d) were likely due to high dissolved organic material (DOM) loading by the swarm. We suspect that mucus shed into the water when jellyfish were concentrated increased the viscosity of the water, and may also have elevated toxins as mucus-bound nematocysts were discharged, as was shown for *S. meleagris* by Shanks and Graham (1988). Qualitatively in support of this, copepods used in a set of egg production measurements from within the swarm were generally lethargic, had increased mortality and produced fewer eggs than copepods outside of the swarm (H. Albright, unpub. data).

Economic considerations

Jellyfish blooms negatively affect economies through effects on tourism, clogging the coolant intake of power generating plants (Rajagopal et al. 1989), interference with diamond mining (J. Field, pers. comm.), fouling coastal fish mariculture pens, consumption of eggs and larvae of commercially important species (Purcell 1985), and reducing fishing effort. We believe the effects on tourism were negligible since *P. punctata*

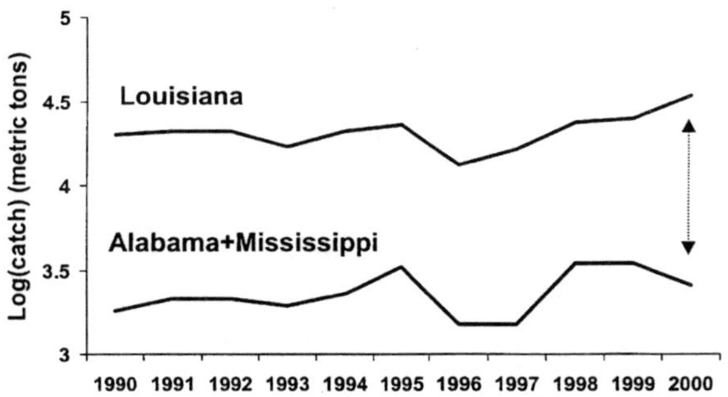

Figure 8. Historical white shrimp (*Penaeus setiferus*) landings, as log(catch), in Louisiana and Mississippi Sound/Mobile Bay regions in the northern Gulf of Mexico. The Mississippi Sound/Mobile Bay region is represented by the sum of Mississippi and Alabama landings. Historical landings of white shrimp in these regions tracked each other up to 2000 when a substantial divergence occurred (arrow).

had a relatively mild sting, and, moreover, the overlap of *P. punctata* with popular tourist beaches in northwest Florida and the northern Mississippi Sound was minimal. In addition, lack of coastal power generators and a mariculture industry focuses negative impacts more on coastal marine natural resources.

The tangible negative economic effects of *P. punctata* include direct effects on fishery harvests, and indirect effects of predation on eggs and larvae of commercially important finfish and shellfish. The Gulf of Mexico commercial shrimp industry is among the most valuable fisheries in the US, and the combined value of commercial shrimping in Alabama and Mississippi in 1990 was $60 million (Weber et al. 1992). For a period of about two months, commercial shrimping effort was substantially reduced, and at times halted, due to fouling of gear and harvest. The most susceptible fishery would have been the coastal harvest of white shrimp (*Penaeus setiferus*), which typically begins in August in the Gulf of Mexico and represents the dominant shrimp fishery of the Mississippi Sound and Mobile Bay. There is circumstantial evidence that *P. punctata* reduced white shrimp harvest in this region during 2000. As presented in Figure 8, inter-annual variations in white shrimp landings in Louisiana and Mississippi Sound/Mobile Bay (represented as the sum of Alabama and Mississippi landings) have historically tracked each other between 1990 and 1999. However in 2000, Mississippi Sound/Mobile Bay white shrimp landings experienced a 26.7% decline, while Louisiana experienced a 37.3% increase (domestic fisheries data from National Marine Fisheries Service, http://www.st.nmfs.gov/commercial). Although a thorough socio-economic study is required to investigate direct and indirect impacts of jellyfish on shrimp harvest, total losses to the commercial shrimp industry due to *P. punctata* interference could have been as high as $10 million ($US) in 2000 (C. Perret, Mississippi Department of Marine Resources, pers. comm.).

The indirect effects on fisheries were less obvious. Consumption of Engraulid-type fish eggs (presumably the anchovy *Anchoa mitchilli*) by *P. punctata* was extraordinary at clearance rates exceeding $90\,m^3\,d^{-1}$ (Table 3). While jellyfish clearance rates on fish eggs are often higher than predation rates on other taxa, the rates we estimated here were higher by at least an order of magnitude than any previously published fish egg clearance rate (see Arai 1997 for a summary). In comparison, Purcell et al. (1994) estimated that the Atlantic sea nettle *Chrysaora quinquecirrha* consumed as much as 29% of the anchovy, *Anchoa mitchelli*, eggs in Chesapeake Bay. The central core aggregation of Lake Borgne, Louisiana covered an area approximating $30\,km^2$ in which the water column was completely processed for fish eggs at least once every day (Table 2). Certainly within the super-swarms these turnover rates were much higher. In Laguna Joyuda, Garcia and Durbin (1993) estimated that *P. punctata* turned over the lagoon zooplankton as much as once every three days.

While consumption of anchovy eggs by *P. punctata* was measurable, the impact on recruitment to the fishery remains intangible. Despite the near-complete consumption of eggs, it was unlikely that the limited *P. punctata* distribution was extensive enough (relative to the total spawning and recruitment area of anchovies in the northern Gulf of Mexico) to reduce recruitment to the anchovy fishery. However, the extraordinary efficiency and magnitude with which *P. punctata* consumed fish eggs does create concern that future blooms might overlap with other specific fish spawning regions such as red drum or red snapper.

Predation on the larvae of commercially important invertebrates by *P. punctata* was variable. While megalopas of portunid blue crabs (*Callinectes* spp.) were recruiting to shallow Mississippi Sound waters in late summer, very few megalopas were recovered from guts of jellyfish. Following the argument that prey swimming velocity must be lower than the marginal bell velocity of *P. punctata* (Costello and Colin 1994, 1995; D'Ambra et al. 2001) we believe that the large, relatively fast megalopas were not being effectively cleared from the water, and it was likely that predation impact on blue crabs was very low. In contrast, predation on larval bivalves, presumably the commercial oyster *Crassostrea virginica*, by *P. punctata* was high (Figure 6, Table 3). Several heavily exploited oyster beds exist in the western Mississippi Sound, and there was high spatial and temporal overlap between *P. punctata* and these beds. Like other rhizostomes, *P. punctata* was an efficient filterer of larval bivalves. Larson (1991) measured high predation rates of larval bivalves by *S. meleagris*, and he estimated that bivalve veligers constituted 47% of the daily ration of *S. meleagris*. However, there was no indication that these high consumption rates translated into larval mortality and reduced recruitment to oyster beds because the mortality of ingested bivalve larvae is uncertain. Purcell et al. (1991) showed that the Atlantic sea nettle *Chrysaora quinquecirrha* ingested, but did not usually digest, oyster larvae in the Chesapeake Bay. Moreover, survival to settlement and metamorphosis remained high for oyster larvae egested after several hours in the guts of *C. quinquecirrha* (Purcell et al. 1991).

Possible fishery for P. punctata

Development of jellyfish fisheries is often considered a means to mitigate the negative effects of jellyfish blooms (Kingsford et al. 2000). Although rhizostome medusae are often a highly prized commodity for Asian markets, it is not likely that *P. punctata* is adequate for fisheries development. Approximately 200 kg of wet biomass of Gulf of Mexico *P. punctata* was processed for human consumption in 2000, however Asian marketing trials revealed that the food-quality of this species of jellyfish was poor (T. Van Devender, pers. comm.). While the water content of *P. punctata* was somewhat higher than other edible Gulf rhizostome *S. meleagris* (Hsieh et al. 1996), it was the presence of the white spots, which created a lumpy processed product that reduced its market appeal (see Kingsford et al. 2000 for a review of jellyfish fisheries).

Understanding spread of P. punctata

If the available evidence suggests that *P. punctata* has been in the Atlantic for nearly half a century, why only now is it invading the Gulf of Mexico? This question is particularly interesting because the very productive Gulf of Mexico, with numerically increasing native jellyfish populations (Graham 2001), would have been as likely 40 years ago to support an invasion of *P. punctata* as in 2000. Carlton (1996) points out several relevant scenarios that are pertinent to the potential success of *P. punctata* as a new invader to the northern Gulf of Mexico. He points out that, while jellyfish are opportunistic animals, invasion success is more than just being 'weedy'. According to the Carlton (1996) model, invasion success of *P. punctata* must be considered in terms of: (1) environmental changes in the donor region (presumably the Caribbean Sea) that has enhanced the supply of medusae, (2) new donor regions within the Caribbean acting as intermediates, (3) environmental changes in the northern Gulf of Mexico that have increased susceptibility to invasion (e.g. increased eutrophication, habitat modification, over-exploitation of fisheries), (4) adequate invasion windows (e.g. physiological tolerances) that promote reproduction and success of the jellyfish population, (5) stochastic factors such as the number and life-history stage of original invaders, and (6) changes in the dispersal vectors.

The introduction of *P. punctata* into the Gulf of Mexico likely represents an inevitable distributional shift of an established Caribbean Sea population that invaded perhaps decades ago. To date we still know very little about the vectors of invasive medusae, and thus any comment on how and when a northern Gulf population was established remains speculative. Since the increased globalization of trade and transportation has been linked to increased rates of exotic species introductions (Carlton and Geller 1993; Ruiz et al. 1997), it is likely that introduction and spread of *P. punctata* has been mediated by open-ocean materials transport. Scyphozoans have a bipartite life-history, in which the medusa represents only the pelagic phase. Bioinvasions by hydrozoans and scyphozoans likely occur at the sessile polyp stage, which presumably arrives in novel environments as part of the fouling community on ships and other sea-going infrastructure (e.g. towed oil and gas platforms), and perhaps not as part of the ballast water community. It was previously assumed that hull-fouling transported *P. punctata* polyps over long distances within the Pacific Ocean and between the Pacific and the Caribbean Sea (Cutress 1971; Larson and Arneson 1990). Transport through the Panama Canal means that polyps were immersed in the freshwater lakes for at least one day. Yet scyphopolyps, which are usually tolerant over wide ranges of salinity, temperature, dissolved oxygen, and food (Arai 1997) could possibly survive such a short duration exposure to freshwater. Moreover, Rippingale and Kelly (1995) experimentally showed that *P. punctata* polyps survived in a state of reduced activity at salinities as low as 15‰.

The initial anecdotal observations of large *P. punctata* (i.e. approximately 35 cm in diameter) in May signified a non-local (i.e. beyond the Mississippi Sound) origin of the population. The coincidence of these medusae with large quantities of the Sargassum weed would suggest that the medusae were advected into the area by an anti-cyclonic spin-off eddy of the Loop Current, which had migrated toward the Louisiana shelf during April 2000 (D. Johnson, NRL Stennis Space Center, pers. comm.). Similar transport of Caribbean medusae had previously occurred (Graham 1998). However, a small population of *P. punctata* is known to have occurred within at least one southern Louisiana embayment for several years prior to 2000 (D. Felder, unpub. data), and a single specimen identified from a fisheries resource survey 70 km south of Louisiana in 1993 (H.M. Perry, unpub. data). Therefore, a plausible alternate hypothesis is that 2000 experienced bloom conditions of a *P. punctata* population that had been cryptic in southern Louisiana, and that advection into Mississippi Sound represented along-shore, rather than cross-shelf, transport perhaps only mediated by the juxtaposition of the anti-cyclonic eddy in the northern Gulf of Mexico.

The overall westward propagation of jellyfish followed the nearshore westerly current of the Mississippi Sound (Figure 3). Though we have no mechanistic explanation at present, the tendency for *P. punctata* to remain nearshore may be a product of behaviors that keep the population entrained in coastal embayments and lagoons in native habitats. The ultimate accumulation of medusae in the far western Mississippi Sound near Lake Borgne was presumably a consequence of the interaction between jellyfish swimming behavior and local hydrodynamics, which are usually responsible for aggregation formation and dispersal (Graham et al. 2001). The ultimate demise of the population in September, though still speculative, was most likely due to senescence of individuals or to increased mortality as storm frequency increased during late summer and fall. We presume this was not atypical timing since similar fall senescence was reported for *P. punctata* in Laguna Joyuda, Puerto Rico (Garcia 1990). *P. punctata* has also showed strong seasonality related to seasonal salinity regime in the Swan-Canning estuary of Western Australia (Rippingale and Kelly 1995).

The ultimate success of *P. punctata* will not be fully recognized until it is known whether future blooms of this species occur. At time of writing this manuscript, we have documented several thousand *P. punctata* along southern Louisiana, west of the Mississippi River. One confirmed sighting has been made east of the Mississippi River near Lake Borgne, Louisiana. Most interestingly, though, is confirmed spread of several hundred medusae in coastal east coast Florida waters in the Indian River Lagoon in 2001 (Figure 1b). This occurrence, though modest by comparison with the northern Gulf of Mexico, has created concern that the species will continue to spread in the region since the highly productive lagoonal habitat of the northern Indian River Lagoon closely resembles the mangrove habitat of *P. punctata* in the western tropical Pacific Ocean. It is not yet known how genetically related the Indian River Lagoon population is to the Gulf of Mexico population. Exactly how reproduction occurs in these jellyfish is an obvious consideration for the success of the species in the Gulf of Mexico. At present, we have no explanation for the lack of females in the population, or whether this population contains a hermaphroditic reproductive strategy. *P. punctata* has always been considered dioeceous (i.e. separate sexes) (Rouse and Pitt 2000), and hermaphroditism in the Scyphozoa is quite rare.

The occurrence and apparent spread of *P. punctata* in the northern Gulf of Mexico and adjacent Atlantic Ocean represents an important and interesting case for understanding jellyfish bioinvasions. We have provided essential information that indicates the greatest impact of these medusae will likely be through food-web controls on zooplankton and through the consumption of fish eggs, and possibly larvae of other commercially important invertebrates such as oysters. Also, direct economic impacts will most likely continue to be experienced in the commercial shrimping industry as these jellyfish interfere with shrimping effort. These medusae appeared suddenly and were substantially different in both size and pigmentation from other populations, which created initial confusion about identification. We anticipate that morphological plasticity is an important characteristic in the ability of this species to successfully invade novel environments. It is hoped, therefore, that the basic ecological understanding we have provided here will be a spring-board to continue investigating this and other jellyfish species in order to understand their invasiveness.

Acknowledgements

This research was funded by the Mississippi-Alabama Sea Grant Consortium (R/CEH-1), the National Sea Grant College (R/CEH-5), and the National Science Foundation (OCE-9733441). We greatly appreciate the efforts of H. Albright, J. Martin, J. Higgins, J. Noel and R. Alley, Marine Technical support at DISL, and the captain and crew of the R/V Pelican in help with field efforts. Two anonymous reviewers provided valuable comments that improved this manuscript. This is Dauphin Island Sea Lab contribution No. 335.

References

Arai MN (1997) A Functional Biology of Scyphozoa. Chapman and Hall, New York, 316 pp

Carlton JT (1996) Pattern, process, and prediction in marine invasion ecology. Biological Conservation 78: 97–106

Carlton JT and Geller JB (1993) Ecological roulette: the global transport of non-indigenous marine organisms. Science 261: 78–82

Clarke TA and Abey GS (1998) The use of small mid-water attraction devices for investigation of the pelagic juveniles of carangid fishes in Kaneohe Bay, Hawaii. Bulletin of Marine Science 62(3): 947–955

Costello JH and Colin SP (1994) Morphology, fluid motion and predation by the scyphomedusa *Aurelia aurita*. Marine Biology 121: 327–334

Costello JH and Colin SP (1995) Flow and feeding by swimming scyphomedusae. Marine Biology 124: 399–406

Cowan Jr JH and Houde ED (1992) Size-dependent predation on marine fish larvae by ctenophores, scyphomedusae, and planktivorous fish. Fisheries and Oceanography 1(2): 113–126

Cutress CE (1971) *Phyllorhiza punctata* in the tropical Atlantic. Proceedings of the Association of Island Marine Laboratories in the Caribbean 9: 14

Dagg MJ (1995) Ingestion of phytoplankton by the micro- and mesozooplankton communities in a productive subtropical estuary. Journal of Plankton Research 17: 845–857

D'Ambra I, Costello JH and Bentivegna F (2001) Flow and prey capture by the scyphomedusa *Phyllorhiza punctata* von Lendenfeld 1884. Hydrobiologia 451: 223–227

da Silveira FL and Cornelius PFS (2000) New observations on medusae (Cnidaria, Scyphozoa, Rhizostomae) from northeast and south Brazil. Acta Biologica Leopoldensia 22: 9–18

deLafontaine Y and Leggett WC (1988) Predation by jellyfish on larval fish: an experimental evaluation employing *in situ* enclosures. Canadian Journal of Fisheries Aquatic Science 45: 1173–1190

Devaney, DM and Eldredge LG (1977) Reef and shore fauna of Hawaii. Section 1: Protozoa through Ctenophora. Bishop Museum Special Publication 64: 1–277

Fancett MS (1988) Diet and prey selectivity of scyphomedusae from Port Phillip Bay. Australia, Marine Biology 98: 503–509

Galil BS, Spanier E and Ferguson WW (1990) and The scyphomedusae of the Mediterranean coast of Israel, including two Lesepsian migrants new to the Mediterranean. Zoologische Medelingen 64: 95–105

Garcia JR (1990) Population dynamics and production of *Phyllorhiza punctata* (Cnidaria: Scyphozoa) in Laguna Joyuda, Puerto Rico. Marine Ecology Progress Series 64: 243–251

Garcia JR and Durbin E (1993) Zooplanktivorous predation by large scyphomedusae *Phyllorhiza punctata* (Cnidaria: Scyphozoa) in Laguna Joyuda. Journal of Experimental Marine Biology and Ecology 173: 71–93

Graham WM (1998) First report of *Carybdea alata* var. grandis (Reynaud 1830) (Cnidaria: Cubozoa) from the Gulf of Mexico. Science 16: 28–30

Graham WM (2001) Numerical increases and distributional shifts of *Chrysaora quinquecirrha* Desor and *Aurelia aurita* Linné (Cnidaria: Scyphozoa) in the northern Gulf of Mexico. Hydrobiologia 451: 97–111

Graham WM and Kroutil RM (2001) Size-based prey selectivity and dietary shifts in the jellyfish, *Aurelia aurita*. Journal of Plankton Research 23: 67–74

Graham WM, Pagès F and Hamner WM (2001) A physical context for gelatinous zooplankton aggregations: a review. Hydrobiologia 451: 199–212

Heeger T, Piatkowski U and Moeller H (1992) Predation on jellyfish by the cephalopod *Argonauta argo*. Marine Ecology Progress Series 88: 293–296

Hsieh Y, Leong F and Barnes KW (1996) Inorganic constituents in fresh and processed cannonball jellyfish (*Stomolophus meleagris*) Journal of Agriculture and Food Chemistry 44: 3117–3119

Ishii H and Baamstedt U (1998) Food regulation of growth and maturation in a natural population of *Aurelia aurita* (L.). Journal of Plankton Research 20(5): 805–816

Ivanov VP, Kamakin AM, Ushivtzev VB, Shiganova T, Zhukova O, Aladin N, Wilson SI , Harbison GR and Dumont HJ (2000) Invasion of the Caspian Sea by the comb jellyfish *Mnemiopsis leidyi* (Ctenophora). Biological Invasions 2: 255–258

Kideys AE (1994) Recent dramatic changes in the Black Sea ecosystem: the reason for the sharp decline in Turkish anchovy fisheries. Journal of Marine Systems 5: 171–181

Kingsford MJ, Pitt KA and Gillanders BM (2000) Management of jellyfish fisheries, with special reference to the Order Rhizostomeae. Oceanography and Marine Biology Annual Reviews 38: 85–156

Kramp PL (1961) Synopsis of the medusae of the world. Journal of the Marine Biological Associations, UK 40: 1–469

Kramp PL (1970) Zoogeographical studies on Rhizostomeae (Scyphozoa). Videnskabelige Meddelelser fra Dansk Naturhistorisk Forening 133: 7–30

Larson RJ (1986) Water content, organic content, and carbon and nitrogen composition of medusae from the northeast Pacific. Journal of Experimental Marine Biology and Ecology 99: 107–120

Larson RJ (1991) Diet, prey selection and daily ration of *Stomolophus meleagris*, a filter-feeding Scyphomedusa from the northeast Gulf of Mexico. Estuarian, Coastal and Shelf Science 32: 511–525

Larson RJ and Arneson AC (1990) Two medusae new to the coast of California: *Carybdea marsupialis* (Linnaeus, 1758), a cubomedusa and *Phyllorhiza punctata* von Lendenfeld, 1884, a Rhizostome scyphomedusa. Bulletin of the Southern California Academy of Sciences 89(3): 130–136

Lucas CH and Lawes S (1998) Sexual reproduction of the scyphomedusa *Aurelia aurita* in relation to temperature and variable food supply. Marine Biology 131: 629–638

Mayer AG (1910) Medusae of the World. Vol III. The Scyphomedusae, pp 499–735. Carnegie Institution, Washington, DC

Mianzan HW and Cornelius PFS (1999) Cubomedusae and Scyphomedusae. In: Boltovskoy D (ed) South Atlantic Zooplankton, pp 513–559. Backhuys Publishers, Leiden, The Netherlands

Moreira MGBS (1961) Sobre *Mastigias scintillae* sp. Nov. (Scyphomedusae, Rhizostomeae) das costas do Brasil. Bolm Inst Oceanography 11: 5–30

Olesen NJ, Frandsen K and Riisgård HU (1994) Population dynamics, growth and energetics of jellyfish *Aurelia aurita* in a shallow fjord. Marine Ecology Progress Series 105: 9–18

Purcell JE (1985) Predation on fish eggs and larvae by pelagic cnidarians and ctenophores. Bulletin of Marine Science 37(2): 739–755

Purcell JE (1989) Predation on fish larvae and eggs by the hydromdeusa *Aequorea victoria* at a herring spawning ground in British Columbia. Canadian Journal of Fisheries and Aquatic Science 46: 1415–1427

Purcell JE (1997) Pelagic cnidarians and ctenophores as predators: selective predation, feeding rates, and effects on prey populations. Annals of the Institute of Oceanography 73(2): 125–137

Purcell JE and Arai MN (2001) Interactions of pelagic cnidarians and ctenophores with fish: a review. Hydrobiologia 451: 27–44

Purcell JE, Cresswell FP, Cargo DG and Kennedy VS (1991) Differential ingestion and digestion of bivalve larvae by the scyphozoan *Chrysaora quinquecirrha* and the ctenophore *Mnemiopsis leidyi*. Biological Bulletin 180: 103–111

Purcell JE, Nemazie DA, Dorsey SE, Houde ED and Gamble JC (1994) Predation mortality of bay anchovy *Anchoa mitchilli* eggs and larvae due to scyphomedusae and ctenophores in Chesapeake Bay. Marine Ecology Progress Series 114: 47–58

Purcell JE, Shiganova TA, Decker MB and Houde ED (2001) The ctenophore *Mnemiopsis* in native and exotic habitats: US estuaries *versus* the Black Sea basin. Hydrobiologia 451: 145–176

Rajagopal S, Nair NVK and Azariah J (1989) Some observations on the problem of jelly fish ingress in a power station cooling system at Kalpakkam, east coast of India. Mahasagar 22(4): 151–158

Rippingale RJ and Kelly SJ (1995) Reproduction and survival of *Phyllorhiza punctata* (Cnidaria: Rhizostomeae) in a seasonally fluctuating salinity regime in western Australian. Marine and Freshwater Research 46: 1145–1151

Rouse GW and Pitt K (2000) Ultrastructure of the sperm of *Catostylus mosaicus* and *Phyllorhiza punctata* (Scyphozoa, Cnidaria): Implications for sperm terminology and the inference of reproductive mechanisms. Invertebrate Reproduction and Development 38: 23–34

Ruiz GM, Carlton JT, Grosholz ED and Hines AH (1997) Global invasions of marine and estuarine habitats by non-indigenous species: mechanisms, extent, and consequences. American Zoologist 37: 621–632

Shanks AL and Graham WM (1988) Chemical defense in a scyphomedusa. Marine Ecology Progress Series 45: 81–86

Shiganova TA (1998) Invasion of the Black Sea by the ctenophore *Mnemiopsis leidyi* and recent changes in pelagic community structure. Fisheries Oceanography 7(3/4): 305–310

Shigonova TA and Bulgakova YV (2000) Effects of gelatinous plankton on Black Sea and Sea of Azov fish and their food resources. ICES Journal of Marine Science 57: 641–648

Strom SL and Strom MW (1996) Microplankton growth, grazing, and community structure in the northern Gulf of Mexico. Marine Ecology Progress Series 130: 229–240

Stoecker DK and Govoni JJ (1984) Food selection by young larval gulf menhaden (*Brevoortia patronus*) Marine Biology 80: 299–306

Suchman CL and Sullivan BK (1988) Vulnerability of the copepod *Acartia tonsa* to predation by the scyphomedusa *Chrysaora quinquecirrha*: effect of prey size and behavior. Marine Biology 132(2): 237–245

von Lendenfeld R (1884) Über eine Übergangsform zwischen Semostomen und Rhizostomen. Zoologischer Anzeiger 5: 380–383

Weber M, Townsend RT and Bierce R (1992) Environmental quality in the gulf of mexico: a citizen's guide. Center for Marine Conservation, Washington, DC, 132 pp

Biological Invasions **5:** 71–84, 2003.

Rapid evolution of an established feral tilapia (*Oreochromis* spp.): the need to incorporate invasion science into regulatory structures

Barry A. Costa-Pierce
Rhode Island Sea Grant College Program, Graduate School of Oceanography, Department of Fisheries, Animal and Veterinary Science, University of Rhode Island, Narragansett, RI 02882-1197, USA (e-mail: bcp@gso.uri.edu; fax: +1-401-789-8340)

Received 17 July 2001; accepted in revised form 4 April 2002

Key words: California, divergence, DNA markers, evolution, invasion histories, regulations for exotic species, tilapia, tilapiine fishes

Abstract

Outside of Asia exotic tilapiine fishes (Trewavas 1983) were not imported directly as native genetic resources from Africa but arrived as transits from third or fourth party sources. Founder populations of exotic tilapia species may be morphologically and meristically distinct in Africa but are still reproductively compatible due to their relatively recent divergence. As a result, feral tilapias have hybridized and introgressed in aquaculture settings before escaping to the wild. Reproductively viable hybrids have resulted, making the use of conventional systematics based upon external morphometric characterizations for species determinations useless. Microsatellite DNA marker studies of 139 tilapia from 10 locations (6 feral, 4 in culture) in southern California, USA, were conducted. Genetic similarities to a worldwide tilapia genetic database were compared by formulating a neighbor-joining dendrogram. The hypothesis that the pattern found arose by chance was tested by bootstrap resampling of 4 microsatellite loci at the 95% level of significance. A significant number of bootstrap re-samples showed that a tilapia species in aquaculture and a feral Colorado River tilapia population were 'monophyletic', meaning they originated from a single source relative to the total variation in the data. A significant number of bootstrap re-samples grouped these two populations with the reference populations of *Oreochromis niloticus* taken worldwide. This 'niloticus' group was found significantly distinct from a second large grouping that included all of the other California tilapia samples and a large group of *O. mossambicus* reference samples taken from a worldwide database. Any regulatory structure attempting to control movements of exotic tilapias based upon discerning 'species' using morphometric and meristic measurements of tilapias is inadequate. California, for example, does not permit *O. niloticus* for aquaculture, but the genetic signature for this species exists in feral stocks collected from the wild in California. There are likely many other places where a certain tilapia 'species' are not permitted but the genetic material exists in established wild stocks within its political jurisdiction. It is recommended that a system of ecotypes (code names) based upon presence of unique DNA microsatellite markers be developed to label and regulate feral tilapia strains, and that hybrid strains be collected into a 'registry' of species based upon unique DNA markers. Such a registry could be used by regulators to better manage exotic tilapias and aquaculture developments.

Introduction

The tilapiine fishes all originate in Africa (Trewavas 1983; Pullin 1988). From the Second World War (WWII) to the present they were exported from Africa – first to Asia – then onwards throughout the world, for the biocontrol of aquatic weeds and insects, and more recently for accelerated tilapia aquaculture developments. Today, the tilapiine fishes are likely the world's most widely distributed exotic fish species, having invaded every tropical and subtropical environment to which they have gained access. This widespread seeding of the world's warmwater ecosystems with the tilapias is similar to the 'carp craze' of the late 19th century in North America (Costa-Pierce et al. 1993). The most widely dispersed tilapia species has been the Mozambique tilapia (*Oreochromis mossambicus*), which was once known as the 'Java' tilapia since most introductions of this fish originated from West Java, Indonesia, its first established locale outside Africa (Hickling 1960). Due to the small sizes of founder stocks, by the mid-1970s the Mozambique tilapia deteriorated in many recipient environments, and small sized, poor quality fish lost consumer acceptance (Pullin 1988).

In the 1970s there was debate among ichthyologists over the systematics of the tilapias. Confusion was due to publication of a proposed new classification system (Trewavas 1973) which split the genus *Tilapia* in two: *Sarotherodon* (the mouthbrooders) and *Tilapia* (the substrate spawners). Trewavas (1982) further revised the tilapias by splitting the mouthbrooding tilapias into two genera, *Sarotherodon* (paternal or biparental mouthbrooders) and creating a new genus, *Oreochromis* (maternal mouththbrooders). Taxonomic confusion occurs in the scientific literature from this period, and is still present among applied workers, especially fisheries and aquaculture scientists and farmers. Fortunately, there have been no major taxonomic revisions of the tilapias since 1982. Trewavas (1973, 1982) revisions have not been accepted by some ichthyologists and fish biologists (the 'lumpers'), nor have they been accepted by the American Fisheries Society (AFS). AFS scientific publications still refer to all of the tilapias as belonging to the single genus of *Tilapia* (Robins et al. 1991). Outside the USA, Trewavas revisions are recognized widely, especially since the publication of a comprehensive monograph (Trewavas 1983). The Trewavas classification scheme has been shown to be inclusive and distinctive (Pullin 1988). For their common name, however, all tilapias may be referred to colloquially as 'tilapias', with a small 't' and no italics (Trewavas 1982), in order to distinguish the common name from the substrate spawning genus *Tilapia*.

Since the 1980s almost all of the worldwide introductions of tilapiine fishes have been for new aquaculture developments. There is no further use of these fishes for biological control purposes. Development of tilapia aquaculture has increased dramatically. Total world tilapia landings from capture and culture has been estimated at 1.16 million t (FAO 1997), with cultured tilapia accounting for 57% of the total (659,000 t). The most important tilapias in aquaculture are the maternal mouthbrooders (Schoenen 1982; Pullin 1985): the Nile tilapia (*O. niloticus*); Mozambique tilapia (*O. mossambicus*); and the blue tilapia (*O. aureus*); plus a number of mouthbrooding tilapia hybrids used in aquaculture (especially red *O. mossambicus* hybrids) with *O. aureus*, *O. niloticus*, and *O. urolepis hornorum*. These species account for 99.5% of global tilapia production (FAO 1997). Of secondary (localized) importance are *Sarotherodon galilaeus* and *S. melanotheron* (principally in West African lagoons). Since the mid-1980s, there has been a shift in producer preferences away from the Mozambique tilapia towards growing Nile tilapia. Nile tilapia now dominates global tilapia aquaculture, accounting for 72% or 474,000 t in 1995 (FAO 1997). Cuba is the world's largest producer of blue tilapia (*O. aureus*), which are grown in an enhanced reservoir fishery supplemented by hatcheries (Fonticiella and Sonesten 2000).

The largest tilapia producing nations are in Asia. China is the world's largest tilapia producer (315,000 t), accounting for 48% of global production, followed by the Philippines, Thailand, Indonesia, and Egypt (FAO 1997). The USA is the world's largest tilapia consumer. US tilapia consumption is estimated at 51,645 t of live weight equivalent (live weight equivalent is calculated as 1.1× the weight of frozen fish and 3× fillet weights) (Engle 1997). The USA imports over 3× the amount of tilapia it grows, with the major importers (in order by value): China, Thailand, Costa Rica, Indonesia, and Columbia (ATA 2000). Tilapia imports contribute a measurable share of the large US trade deficit in seafood products. Tilapia are the third largest imported aquaculture product to the US after shrimp and salmon (ATA 2000). Tilapia have received rapid consumer acceptance in US seafood circles as the 'new white fish'. Tilapia are the most frequently requested fish in the US restaurant trade, and new markets carrying the fish for the first time report rapid acceptance (ATA 1995). The culinary characteristics of the fish match almost perfectly the desires of the US consumer, for example, a white flesh, boneless, relatively odorless, with a very mild flavor. Tilapia are increasingly being seen as a replacement for cod and hake which are in short supply. Tilapia sales have exceeded those of trout in the US each year since 1995 (ATA 2000). As a result, tilapia production in the Americas is expected to exceed 500,000 t by 2010 (Fitzsimmons 2000).

Invasion histories of tilapiine fishes: a case study of California, USA

Since tilapia have established in nearly every water body to which they have had access worldwide, a case study of the tilapias in a US state that prohibits certain tilapia species was undertaken to discern if tilapia species regulations were congruent with current status of stocks in the wild and in captivity in selected farms. California state regulations permit aquaculture of three tilapiine fishes (*O. mossambicus, O. u. hornorum,* and *Tilapia zillii*), plus restrict tilapia aquaculture to the southern part of the State below the Tehacapi Mountains in the counties of San Bernardino, Los Angeles, Orange, Riverside, San Diego, and Imperial. Other tilapia species used widely for aquaculture worldwide, especially the Nile and blue tilapias (*O. niloticus* and *O. aureus*), are illegal. California shares the Colorado River with Arizona which permits the importation and culture of Nile tilapia. In addition, the California Department of Fish and Game (CDGF) issued – for the first time – a permit to one farm to grow Nile tilapia in a closed, indoor recirculating system where all fish are marketed as processed products (e.g. no live fish sales are permitted).

*Zill's tilapia (*Tilapia zillii)

T. zillii are native to a large swath of north central sub-Saharan Africa from Senegal in West Africa through northern Zaire and the Sudan, and north into the Nile River basin and Asia Minor (Pullin 1988). Its distribution extends south to the central African rainforest and around Kisangani, Zaire where it meets the distribution of another closely-related substrate spawner, *T. rendalli*. It is believed that *T. zillii* and *T. rendalli* were the same species until recent time when the 'drying' of Africa occurred and they became separated into a 'northern' form (*T. zillii*) and a 'southern' form (*T. rendalli*) (Trewavas 1983).

T. zillii were imported to southern California due to their ability to feed on nuisance aquatic weeds and other macrophytes which were clogging irrigation canals. It was hoped *T. zillii* could be a biological control agent to offset the high costs of mechanical and chemical controls (Hauser 1975a, b). *T. zillii* was first imported to California (University of California Davis and University of California Riverside, UCR) in the 1960s from the Arizona Cooperative Fishery Unit (ACFU), University of Arizona, but these died out (Legner and Pelsue 1977). Another permit was issued on October 15, 1971 to import 150 *T. zillii* from the ACFU to the Division of Biological Control, UCR. Legner and Fisher (1980) reported the *T. zillii* in Arizona originated from just three male/female pairs imported from Israel in 1965.

T. zillii are noted for their hardiness, having wide temperature (7–42 °C) and salinity tolerances (upwards of 45 ppt; Chervinski 1971). However, Hauser (1977) found that at water temperatures below 16 °C *T. zillii* became lethargic, lost equilibrium, and were vulnerable to predators and disease. In irrigation canals in Imperial County, the fish were stressed frequently by low temperature from mid-December to March, and mortalities were observed in December–January (Hauser 1974, 1977). Survival of *T. zillii* in southern California occurred only in areas where warm seepage water or geothermal waters created thermal refugia for overwintering (Hauser 1977).

For both biological control and aquaculture purposes *T. zillii* is a poor choice because of its high fecundity and high spawning periodicity; its slow overall growth rate to a small maximum size; and its narrow temperature optimum for good growth. However, Hauser (1975c) reported an unusually fast growth rate of *T. zillii* from a California irrigation ditch (the highest ever recorded; see Pauly et al. 1988). Two-year-old fish reached 380–709 g. All other reports of growth are much slower (Pullin 1986). Fecundity in *T. zillii* is 10–20 times higher than the mouthbrooding tilapias. Lowe-McConnell (1955) reported that 10 mixed sex *T. zillii* of just 2.6–4.8 cm total length produced 9000 progeny in 9 months. Platt and Hauser (1978) found that at water temperatures less than 20 °C, feeding rates on macrophytes and growth rates of *T. zillii* approached zero. Hauser (1977) concluded that due to its thermophilic nature, *T. zillii* would not be self-sustaining in southern California irrigation ditches and canals without thermal refugia, and that any effective biological control of aquatic weeds in irrigation canals would require expensive annual restocking.

T. zillii have been implicated in the decline of the desert pupfish in the Salton Sea area (Schoenherr 1988). In addition, *T. zillii* have been collected from southern California coastal waters near a power plant off Huntington Beach, and in Upper Newport Bay in Orange County (Knaggs 1976). No known marine collections have been made more recently.

T. zillii populations in southern California have a restricted genetic basis (the progeny of just 3 reproductive pairs were imported to Arizona and distributed to California), presenting severe genetic bottlenecks and restricting the adaptability of the species. The species has poor environmental compatibility with California: it is of no use as a cost-effective biological control agent; and it has no economic value to aquaculture, capture, or recreational fisheries. *T. zillii* can adapt to natural and man-made thermal refugia; it impacts aquatic vegetation during warm months of the year, and its populations reportedly conflict with restoration programs for native, desert fishes. It is predicted that introduced populations of *T. zillii* will decline in California and will eventually disappear altogether. It is recommended that the decline of *T. zillii* be studied for the insights it may give to the eradication of other nuisance exotic aquatic species; and that studies be made on cost-effective, ecological approaches to hasten its demise and rapid eradication from California waters.

The Mozambique tilapia
*(*Oreochromis mossambicus)

The Mozambique tilapia is native to the eastward-flowing rivers of east Central Africa. In the northern part of its range it is present below Kapachera Falls in the lower Shire River in southern Malawi, the lower Zambezi, and in Mozambique in all coastal rivers down the southeastern African coast to Algoa Bay, South Africa (Pullin 1988).

The Mozambique tilapia and the common carp are likely the two most widely spread exotic aquatic animals in the World. From the 1930s to 1970s the Mozambique tilapia was spread throughout the world by fisheries biologists and managers for the control of nuisance aquatic weeds and insect pests, and for aquaculture and fisheries development. The Mozambique tilapia is more widely farmed in Asia and elsewhere than in its genetic home of Africa. The route *O. mossambicus* first took out of Africa to Asia is unknown. Tilapia were 'discovered' in Asia – first in Java, Indonesia in 1938 – and were hailed as a miracle (Schuster 1952). Some scientists believe the Indonesian tilapia came from transfer by an aquarist in Singapore in 1938 (R.S.V. Pullin, pers. comm.). During WWII they served as invaluable protein food for the poorest of poor ravaged by Japanese occupation. Dr Schuster, a Dutch fisheries biologist, reported a fish farmer in Java found five tilapia (two females and three males) in a coastal brackishwater pond (Schuster 1952). After WWII, Schuster returned to Indonesia (he was interned as a prisoner of war by the Japanese) and found that *O. mossambicus* had spread throughout Java.

Hawaii was the first state in the USA to receive tilapia from Asia. In 1951, 60 'small' Mozambique tilapia (then known as 'Java tilapia') were sent to Honolulu from Singapore (Brock 1960). The Singapore tilapia is of unknown origin but likely resulted from wartime traffic across the Straits of Malacca (Singapore and Indonesia). Only 14 Mozambique tilapia survived transit to Hawaii from Singapore (unknown numbers of males and females) (Brock 1960). These fish were bred and 'the resultant offspring were used for stocking purposes' in Hawaii and elsewhere in the Pacific (Brock 1960). Because of the unknown origin and numbers of the founder stocks sent from Africa to Asia, Pullin (1988) has raised the specter that *all* Asian *O. mossambicus* populations could be derived from the five fish found by Schuster in Java. And Asia – not Africa – is the origin of all of the feral *O. mossambicus* established throughout the world.

In the USA, Arizona imported its *O. mossambicus* stock from Hawaii in early 1961. By October 1961, more than 27,000 progeny of the original Hawaii fish were produced. These were stocked into the Gila Bend Canal. In January 1963, an unknown number of fish were released into the Yuma Canal (Barrett 1983). In 1962, McConnell (1966) began experiments in Arizona with the 'Malacca hybrids' importing a second stock of Mozambique tilapia from the Tishomingo, Oklahoma National Fish Hatchery, and also bringing an unknown number of *O. u. hornorum* from Malacca, Malaysia (Hickling 1960). Reproductively viable hybrids of these two species were released widely to canals, dams, ponds, and reservoirs in Arizona from 1965 to 1981 (Barrett 1983).

Interestingly, the 'Tishomingo' strain of *O. mossambicus* appears to have originated in California! The Oklahoma strain was imported to Tishomingo from Auburn University. Auburn originally obtained its Mozambique tilapia from the Steinhart Aquarium (San Francisco, California) in 1953 (Keely 1957). Steinhart had earlier obtained their stock from Hawaii.

Hickling (1960) pioneered the 'all male hybrid progeny' cross of a male *O. u. hornorum* (then called the

'Zanzibar tilapia') and a female *O. mossambicus* ('Java tilapia') which gained worldwide attention. In December 1963, the California Department of Fish and Game (CDFG) obtained '8 males of the Zanzibar strain, plus 50 male and female fingerlings, and 6 adult males of the Java strain' from the ACFU for experimental use at the CDFG station in Chino, California (Outdoor California 1964; St Amant 1966). The fish were authorized for use as a biological control of aquatic weeds and insects in irrigation systems of southern California on November 5, 1971. The Mozambique tilapia and hybrids were officially stocked in southern California thereafter.

By 1968, tilapias had been found in some 15 miles of irrigation canals (the Araz Drain and Reservation Main Drain) near Bard in Imperial County, CA (Hoover and St Amant 1970). Since tilapia were not officially stocked into southern California waters until 1971 (Pelzman 1973), these tilapias likely represented movements into California from 1961 to 1963 stockings of Arizona tilapias in irrigation canals connected to California.

In a separate development, in January 1964, Mozambique tilapia were discovered illegally in 0.25 acre pond at a private tropical fish farm near the 'Hot Mineral Spa' (Niland, California) (St Amant 1966). The pond was poisoned with rotenone and over 5000 tilapia removed by the CDFG. However, in June 1965, CDFG workers observed a breeding population had re-established in a drainage ditch choked with emergent macrophytes that was connected to the Salton Sea. A enormous and self-sustaining hybrid tilapia population exists in the Salton Sea (Costa-Pierce and Riedel 2000).

The Mozambique tilapia has penetrated California's coastal marine waters but does not appear to have established. Horn (1988) reported that the tilapias are 'now part of the Upper Newport Bay fauna'; however, Horn collected no tilapia in a year-long (1978–1979) study. Tilapia were, however, observed to build nests at the mouth of San Gabriel River in other years. Fishers at 'Warmwater Beach' off the Encino Power Plant in Carlsbad, CA reported catching tilapia in the nearshore warmwater discharge (Costa-Pierce, pers. obs.). Apparently, where thermal refugia exist along the southern California coast, the Mozambique tilapia have been able to penetrate seawater. The biology and environmental impacts of these marine tilapia populations are unknown: if these populations are fully saline; if spawning occurs in freshwater; or if fish are simply flushed to the coastal areas during the rainy season and then adapt to marine conditions.

O. mossambicus is most saline tolerant of the tilapiine fishes, having evolved from a marine ancestor (Myers 1938; Whitfield and Blaber 1979; Trewavas 1983; Costa-Pierce and Riedel 2000). *O. mossambicus* adapts especially well to environments where salinities change gradually (Philippart and Ruwet 1982). For example, in St Lucia Lake, a coastal lagoon along the southeast coast of Africa, irregular tidal connections lead to a slow sequence of low to high salinities, and *O. mossambicus* adapts well to salinities from 0 to 120 ppt. Increased salinities, however, may cause rapid reductions in biomass. *O. mossambicus* was 12% of the catch at 10 ppt, dropping to 1% at 80 ppt (Whitfield and Blaber 1979). *O. mossambicus* hybrids have a higher growth rate in seawater than freshwater due to an increased rates of food consumption as salinities increase (Watanabe et al. 1988). Reduction in the rate of routine metabolism may account for the increased growth rates of tilapia in seawater over freshwater animals (Ron et al. 1995).

O. mossambicus are euryhaline, but stenothermal, dying when water temperatures fall below 5–10 °C (Mironova 1969; Trewavas 1983; Wohlfarth and Hulata 1983; Costa-Pierce and Riedel 2000). Lower water temperatures decrease growth rates and halt reproduction. Laboratory experiments showed a sharp depression of growth for tilapia kept at 20–22 °C compared with tilapia at 25–28 °C (Chmilevskii 1998).

Small sized founder stocks, resulting genetic bottlenecks and introgression with *O. u. hornorum* put the species designation of feral *O. mossambicus* as a distinct species in southern California (and in many similar areas of the world) into question. Like most tilapias, the population genetics of native Mozambique tilapia from Africa has been little studied. Lombard (1960) examined the culture performance of some *O. mossambicus* strains in Africa and found a wide diversity of growth to larger maximum sizes than reported in the modern tilapia aquaculture literature. Caulton (in Pullin and Lowe-McConnell 1982, pp 333–334) stated that *O. mossambicus* from the lower/middle Zambezi had a much better appearance than those found elsewhere in the world, and that fish from South Africa performed well in culture.

Mozambique tilapia have negatively impacted many native tropical and subtropical aquatic ecosystems due to their aggressive, broad-spectrum omnivorous feeding habits, and precocious breeding behavior (Lobel

1980). Its species mouthbrooding habitats allow it to care for and carry its young, migrate, and disperse its progeny far from its spawning sites to seed (and impact) new, distant areas far from the site of original introductions. Its trophic plasticity and ability to switch from one prey to another are notable (DeSilva et al. 1984). Mozambique tilapia have even became a predator in some locales (on milkfish fry, Lobel 1980). As a result of these undesirable traits, the Mozambique tilapia has been replaced in nearly all of the major tilapia producing countries by the Nile tilapia (*O. niloticus*). The global consensus that the Nile tilapia (*O. niloticus*) is the best species for aquaculture development is due to at least three important biological characteristics:

- its faster growth rate to larger maximum sizes than other species;
- its larger size at first reproduction;
- its stenophagy, with grazing feeding habits, and consequent lower position on the aquatic food web, making it cheaper to grow, and with less environmental impact (Moreau et al. 1986; Pauly et al. 1988; Pullin 1988).

As a result, from 1984 to 1990 Nile tilapia aquaculture production expanded at an annual percentage rate (APR) of 22.6% while the APR for Mozambique tilapia was 8.9% (FAO 1997).

*The Wami River tilapia (*Oreochromis urolepis hornorum)

The Wami River tilapia originates from the Wami River, in eastern Tanzania. *O. urolepis* has been found in four coastal locations in Tanzania, but the Wami River subspecies, *O. u. hornorum*, is known only from the Wami River and Zanzibar Island (Trewavas 1983). The Wami River tilapia, however, is not native to Zanzibar. An unknown number of fish were imported to Zanzibar from the Tanzanian mainland in 1918 by Dr W.M. Aders for experimental fish culture (Talbot and Newell 1957). It is unknown if the Wami River subspecies was mixed with a native Zanzibar tilapia before its worldwide distribution.

The Wami River tilapia is famous in aquaculture not for its role as a distinct species grown for its own merits but as the male parental stock with female *O. mossambicus* used to produce 'all male' hybrid progeny (Hickling 1960). This development elicited worldwide interest; as a result, the Wami River tilapia was exported worldwide for hybridization purposes. Hickling (1960) reported receiving only 36 fingerling Wami River tilapia from Zanzibar at the Fish Culture Research Station, Malacca, Malaysia. Therefore, Hickling's and the experiments of other workers worldwide may have been affected by the small size of the original founder stocks and the fact that the founder stock came not from its place of origin but from a fish farm in an exotic location that may have already been hybridized and introgressed.

McConnell (1965, 1966) began experimentation with the 'Malacca hybrid' at the ACFU by importing in 1962 an unknown number of Wami River tilapia (*O. u. hornorum*) from Malaysia. In Arizona, there were few controls taken to prevent mixing of parental stocks from the reproductively-viable hybrid progeny. As a result, uncontrolled hybridization and backcrosses occurred (Barrett 1983). Barrett (1983, p. 5) stated, 'Both parental species and reciprocal hybrids were introduced into ponds to assess their survivorship and angler susceptibility.'

There is confusion about the status of the Wami River tilapia in brackish and marine waters in southern California. Legner and Pelsue (1977) refer to large reproducing populations of '*T. hornorum*' in Coyote Creek, Los Angeles due to continuous power plant waste heat discharge into the creek at 85 °F year round. He states, 'Daily tidal washes of the site favor the persistence of this fish over the other two species (e.g. *O. mossambicus* and *T. zillii*) which are less interhaline (sic).' Talbot and Newell (1957) reported *O. u. hornorum* grow and reproduce in 33–39 ppt water. One aquaculture company in Niland, CA has selected tilapias from the Salton Sea for broodstock and reports persistence of a grayish–brownish tilapia fitting the external characterization of *O. u. hornorum* according to Trewavas (1983); however, no biochemical work has been done to confirm that a distinct genetic signature of *O. u. hornorum* persists in southern California. Hybrids of *O. mossambicus* × *O. u. hornorum* were stocked extensively in California to the point that it is unlikely that pure lines of *O. u. hornorum* (the Wami River tilapia) or *O. mossambicus* exist. Strangely, the Wami River tilapia itself still has unknown aquaculture potential since it is nowhere in the world grown as a pure species.

The failure of systematics to discern introduced tilapias

Studies of feral tilapias have been conducted using systematic (taxonomic) nomenclature based upon external color and appearance, distinctive spawning behaviors, and morphometric and meristic measurements (Minckley 1973; Barrett 1983; Lopez and Ulmer 1983). Barrett (1983) completed an extensive morphometric and meristic study of tilapias in the lower Colorado River basin. He reported that *O. mossambicus* and *O. u. hornorum* have hybridized to the extent that they are essentially one reproducing population. He also reported that *O. aureus* (the blue tilapia) was the 'dominant cichlid in Arizona' and that 'recent evidence indicated that *O. niloticus* (the Nile tilapia) genes were also present.'

Taxonomic studies of introduced tilapias using external characteristics cannot resolve tilapia species differences because of the 'supraspecies' nature of the tilapias (Pullin 1988). Tilapias in Africa have a wide distribution considered ancient. Due to the 'recent' drying of the continent, tilapia populations with wide distributions have become isolated. 'Newly' separated populations have evolved rapidly into new species through new mutations and genetic flux.

The concept of the 'supraspecies' is where a widely distributed species is 'falling apart' (Pullin 1988). The 'newly' isolated populations do not freely interbreed unless brought together by some major environmental event (such as floods, earthquakes, etc.). Recognition of the level of differentiation necessary to assign a given population to a new 'species' is a matter of expert opinion – the classic 'lumpers' *versus* 'splitters' debate (Pullin 1988).

In aquaculture settings, or anywhere in the tropical aquatic environment where large scale introductions of exotic tilapias occur, there is a risk of interbreeding and hybridization among populations that may be morphologically and meristically distinct in Africa but reproductively compatible when put together as exotic species due to their relatively 'recent' divergence. Where a mixture of tilapia species have been stocked, reproductively viable hybrids have resulted, and the use of external characterizations of these hybrids for species determinations is fruitless (Wohlfarth and Hulata 1983; Pullin 1988).

Use of genetic markers to identify and classify feral tilapias

Genetic investigations of tilapias in southern California were based upon use of 'microsatellite' loci (Costa-Pierce and Doyle 1997). These loci consist of variable numbers of tandemly-arranged repeats of simple base sequences (VNTRs). There are a number of reasons why microsatellite polymorphisms have become the markers of choice for fisheries genetic studies (Wright and Bentzen 1994; Bert et al. 2002):

1. the level of polymorphisms at microsatellite loci is usually very high – heterozygosities approach 100% at some loci – allowing for identification of families in hatcheries and high resolution population and species identifications;
2. the alleles at each locus are inherited in strictly co-dominant Mendelian fashion (except in cases of high mutation rates) which facilitates analysis of hybrid populations at the second and sebsequent generations;
3. the scoring procedure is numerical and gene frequency information can be stored, compared, and exchanged among researchers working on different populations and species in different laboratories without ambiguity;
4. numerical scoring permits semi-automated data collection on automated sequencing equipment, giving rise to attractive economies of scale;
5. microsatellite systems developed for one species frequently work for other closely-related species, greatly reducing effort required to adapt procedures to new problems and resolving the composition of hybrid species;
6. microsatellite polymorphism has been recommended by an FAO Working Group as the preferred genetic identification system for a global program on conservation and identification of breeds of domestic livestock (Barker 1994). If adopted, technologies developed (including statistical and data processing analyses) for terrestrial and aquatic species will be compatible and will develop rapidly.

Other DNA-level polymorphic systems in use in fisheries and aquaculture research are restriction fragment length polymorphisms (RFLP) in non-repetitive (e.g. single copy) nuclear DNA, multi-locus minisatellite DNA fingerprinting, randomly amplified polymorphic DNA (RAPD), and mitochondrial DNA RFLP and sequence polymorphisms. None of these systems combines all of the favorable features of microsatellites (Wright and Bentzen 1994). Protein polymorphisms (mainly isozymes or allozymes) have also been used to mark or identify genotypes in aquaculture and for breed comparison and genetic stock identification (GSI). GSI is the analysis of the composition of mixed stocks; that is, stocks composed of mixtures of fish of different origins (Millar 1987; Brodziak et al. 1992).

The laboratory practice and underlying molecular biology of DNA microsatellite polymorphism in fish has been comprehensively reviewed (Franck et al. 1991; Wright and Bentzen 1994). The basis of the polymorphism is variation in the number of copies of tandemly-repeated sequences of 2–6 bases (GCGCGC..., ATAATAATA..., ATAGATAGATAG..., etc.). Changes in the copy number are thought to occur during meiosis through slipped-strand mispairing or slippage of DNA during replication (Schlotter and Tautz 1992). VNTRs with 2 bp repeats, called dinucleotide microsatellites, have been used in this study.

The polymerase chain reaction (PCR) provides a simple and relatively unambiguous detection procedure for microsatellite polymorphisms when the objective is pedigree analysis of large numbers of fish. The PCR procedure involves the synthesis of pairs of oligonucleotides (primers) that are complementary to unique sequence DNA lying adjacent to each end of the variable length VNTR locus. In manual scoring procedures the intervening VNTR sequence is amplified by PCR using radioactive primers, electrophoresed on acrylamide gels, and scored using standard autoradiographed procedures. High-resolution sequencing gels are required to avoid confounding alleles which differ by only 2 bp (dinucleotide repeats). The amplified PCR products separated by electrophoresis can also be detected by automated flurometric procedures (Ziegle et al. 1992; Kimpton et al. 1993). Tetranucleotide repeats can be resolved non-radioactively on simpler agarose systems but appropriate primers are more difficult to isolate. Tetranucleotide repeat systems have been developed for rainbow trout (Herbinger et al. 1995). More efficient procedures for identifying microsatellite primers are becoming available. Santibanez-Koref et al. (1993) used PCR determination of flanking sequences of microsatellites in a procedure which involves initial cloning into plasmids, lambda phage, or yeast chromosome vectors, followed by digestion with restriction enzymes, ligation to appropriate linkers, and final amplification with primers complementary to the microsatellites and linker sequences.

DNA marker technology (microsatellite DNA polymorphism) was used, and two types of statistical analyses conducted with tilapias collected from southern California. DNA markers were compared with reference DNA markers in the tilapia database at the Marine Gene Probe Laboratory, Dalhousie University, Halifax, Nova Scotia, Canada. Two hypotheses were tested on the DNA markers: (1) that none of the alleles in California tilapia samples came from *O. niloticus*; (2) that all or most of the microsatellite alleles came from *O. mossambicus*.

Sample collection, preservation, and processing

Ten different tilapia samples existing either in the wild (e.g. feral) or grown commercially were obtained; 139 fish were collected from six locations in the wild and four tilapias grown in aquaculture. Samples of fish collected and analyzed are given in Table 1.

Fish samples were obtained either by use of a gill net, seines, or hand nets from tanks and ponds. Fresh fish were wrapped individually in foil and put in ziplock bags on ice in the field, then frozen at −20 °C upon return to the laboratory. Fish were defrosted and 0.5–1.0 g pieces of muscle tissue without skin from the middle of the dorsal area above the lateral line were taken and put into 1 ml of 95% ethanol (completely denatured) in 2.0 ml screwcap Eppendorf microcentrifuge tubes.

DNA was extracted from muscle tissue according to the procedures of Ruzzante et al. (1996). This procedure (with some modification) can also be used to extract DNA from blood or fin clips taken as non-lethal samples for broodstock verification (Brooker et al. 1994; Herbinger et al. 1995; McConnel et al. 1995). A sample of muscle approximately 100 mg was washed with 100 mM Tris-HCl and 40 mM EDTA pH 8.0 to remove the ethanol and placed in 500 μl of lysis buffer (100 mM Tris-HCl, 100 mM EDTA, 250 mM NaCl, 1% SDS at pH 8.0) to which Proteinase K (ICN Biomedicals) was added to a final concentration of 250 μg/ml.

Table 1. Samples of fish analyzed from the field and reference samples.

(1) Salton Sea (8 fish)
(2) Three different outlet ditches draining to Salton Sea from tilapia farms (41 fish)
(3) 'niloticus' population at tilapia farm # 2 (10 fish)
(4) 'mossambicus' population at tilapia farm # 2 (10 fish)
(5) California tilapia farm # 1 (10 fish)
(6) California tilapia farm # 3 (20 fish)
(7) Colorado River, Blythe (20 fish)
(8) San Gabriel River, Los Angeles (20 fish)
(9) *O. niloticus*, ICLARM, Philippines but collected from West Africa (30 fish)
(10) *O. mossambicus*, University of Malawi, Malawi, Africa (36 fish)
(11) *O. mossambicus*, Rhodes University, South Africa (10 fish)
(12) *O. urolepis hornorum*, Wami River, Tanzania (7 fish), Stirling University, UK (2 fish)

The sample was incubated 18 h at 55 °C to digest the tissue. Proteins were removed from the tissue lysate by extracting the sample first with an equal volume of buffer saturated phenol (pH 8.0), transferring the aqueous phase to a clean test tube, then extracting with an equal volume of chloroform (Sambrook et al. 1989). The aqueous phase was again transferred to a clean test tube and DNA precipitation done by adding 2 volumes of 100% ethanol. The mixture was placed at −20 °C for at least 2 h and the precipitated DNA recovered by centrifugation for 5 min. The DNA pellet was washed with 1 ml cold 70% ethanol, spun for 2 min, and the alcohol discarded. The DNA pellet was air dried for 10 min and resuspended in 100 μl of 10 mM Tris-HCl and 0.1 mM EDTA (pH 8.0).

PCR amplification of purified DNA was accomplished by the following modifications of Brooker et al. (1994). The PCR cocktail was: 10 mM Tris-HCl at pH 8.3; 50 mM KCl; 1 mM $MgCl_2$; 0.01% gelatin; 400 μM each dNTP; 0.2 μM each primer; 0.001 μM reverse primer labeled with ^{32}P; 0.25 U Taq polymerase; 25 nG purified DNA. The labeling of primer with ^{32}P consisted of: 100 pM of reverse primer; 50 pM of ATP–g–^{32}P; 1X Pharmacia OnePhorAll buffer; 10 μl Polynucleotide kinase. The cocktail was incubated for 30–60 min at 37 °C and kinase inactivated by heating at 65 °C for 15 min. Amplification was conducted in 5 μl total in a MJ PTC-100, first for 7 cycles of: denaturation at 94 °C for 1 min; primer-specific annealing for 30 s; extension at 72 °C for 30 s. This was followed by 38 cycles of: denaturation at 90 °C for 30 s; primer-specific annealing for 30 s; and extension at 72 °C for 30 s. Following amplification, a stop dye (99% deionized formamide; 10 mM NaOH; 0.1% bromophenol blue; 0.1% xylene cyanol) was added to the sample and it was denatured for 15 min at 94 °C. Then 3.5 μl of the sample/dye mix was electrophoresed through an 8% denaturing acrylamide gel. The gel was run at 50 mA constant current for approximately 3 h, fixed, dried, and exposed to Kodak X-Omat AR5 film at −80 °C for 18–36 h and the products sized relative to an M13 sequence ladder (Ziegle et al. 1992).

The different locus annealing temperatures for the microsatellite loci analyzed were: 57 °C for Os7R; 51 °C for On13 (unpublished); 49 °C for Os75, and 54 °C for both Os64 and Os25. The primer sequences were: Os7R – forward (CCA GGG AGA GGA AAT GAG C) and reverse (GAT GCG GCA ACA GTT ATG TC); Os75 – forward (AGC CTA AAA TAA TGG AAT CAC) and reverse (CCA CAG AGT CAT GGT TCA C); Os64 – forward (CAG TGT CTT CAG TTC CTT GC) and reverse (CAG AAG CAT CTT ATT GAT GAC); Os25 – forward (TTG TGA AAT TGC ATT GCA CTC) and reverse (AAC TCC CTT TGA TCC TCT GC).

Genetic characterization of California tilapias using microsatellite alleles

Average heterozygosities in many of the samples were as high as found at microsatellite loci in natural populations of other fish (Table 2) (Wright and Bentzen 1994; Herbinger et al. 1996). However, the low heterozygosity of the aquafarm 1 sample (reported to be *O. mossambicus* when collected) indicates a strong probability that a serious ‘bottleneck’ or inbreeding has occurred (0.21 vs. a range of 0.34–0.72 for other samples).

Overall genetic similarities among samples and relationships to reference species are illustrated by a neighbor-joining dendrogram (Saitu and Nei 1987) (Figure 1). These dendograms imply no evolutionary relationships but give a visual impression of the way populations are grouped according to Nei’s genetic distance from each other (Nei’s distance adjusted for small sample size; Nei et al. 1983). The relationships among populations shown in Figure 1 are as likely to be affected by mixing and hybridizations as by evolutionary processes leading to divergence.

The hypothesis of the likelihood the pattern shown in Figure 1 arose by chance was tested by bootstrap resampling of the four microsatellite loci (Felsenstein 1985). If a given element of the pattern arises in 95% or more of the random bootstrap samples it can be considered to be a real property of the data and highly significant. It also indicates that the relationships are not highly dependent on only one of the four loci.

A significant number (99%) of bootstrap re-samples of the 4 loci showed that one of the populations sampled at farm #2 (a permitted ‘niloticus’ population) and the

Table 2. Average DNA microsatellite heterozygosities and standard errors for tilapias sampled in southern California.

Place	Heterozygosities
Salton Sea	0.72 ± 0.07
Feral in Salton Sea ditches	0.64 ± 0.12
Feral in Colorado River	0.35 ± 0.16
Aquafarm 1	0.21 ± 0.14
Aquafarm 2	0.53 ± 0.19
Africa 1 reference (Malawi)	0.68 ± 0.06
Africa 2 reference (South Africa)	0.19 ± 0.00

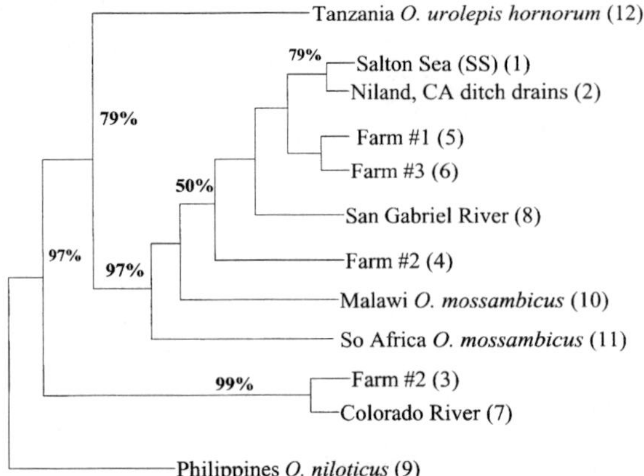

Figure 1. Unweighted pair-group method arithmetic (UPGMA) dendrogram (Saitu and Nei 1987). Sample designations correspond to the numbers in Table 1. Bootstap frequencies are shown as numbers associated with the branches. The 'California Mossambique' ecotype would comprise samples 1, 2, 4, 5, 6, and 8 since this group is separated from other samples by 97% of the bootstap resamplings. The data also indicate that this *O. mossambicus* is a hybrid with *O. u. hornorum*. The 'California niloticus' ecotype comprises samples 3 and 7, are grouped with the *O. niloticus* reference strains, and are separated from other samples 97% of the time.

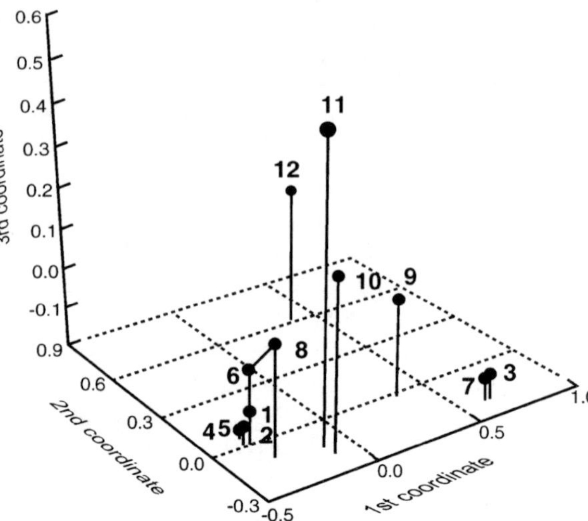

Figure 2. Principle component plot (Cavalli-Sforza et al. 1994) derived from 113 alleles at 4 loci. The three principle coordinates account for 69% of the total variance among groups. The members of each ecotype are linked by minimum spanning trees for ease of visual identification. The 'mossambicus-hornorum' hybrid ecotype is a clearly distinguished grouping (1, 2, 4, 5, 6, 8) from the 'niloticus' ecotype (3, 7). The third coordinate separates the African reference samples (9, 10, 11, 12) from all of the California samples, indicating significant divergence of the established California exotics from native genetic material in Africa.

Colorado River tilapia populations are 'monophyletic'. In this context this means the tilapias come from a single source relative to the total variation in the data. An almost equally large (97%) of re-samples group these two populations with the worldwide reference populations of *O. niloticus.* This first group is distinct (significant at the 95% level) and separate from a second large grouping that includes most of the other samples and the large *O. mossambicus* database taken worldwide (Figure 2).

The existence of the first group shows that *O. niloticus* genetic material is present in samples in California. The second group includes the farm #2 'mossambicus' population, all other samples, plus the *O. mossambicus* reference database. In this second group, the Salton Sea and Niland ditch samples consistently group together during bootstrap re-sampling. However, the internal relationships among the second group are not statistically significant at the 95% level; there is no assurance that this group contains genetic material from only one species; in fact, it is highly likely these fish are hybrids.

These analyses summarize the relationships among southern California tilapias for a single variable, for example, genetic distance, and are shown in only one dimension. This approach is appropriate for the study of evolutionary processes where gene pools (e.g. species) are assumed to diverge independently after speciation. In California, a more interesting question is whether there has been (or might be in the future) mixing of genetic material of different origins. The question is whether California has experienced introductions and possible hybridizations of tilapia species that separated from each other so long ago that their African evolutionary phylogenies at the species level are not relevant. Principal coordinate analysis (PCA), or multi-dimensional scaling, has been successfully employed in analogous situations to trace the evolution, movement, and mixing of human populations (Cavalli-Sforza et al. 1994). PCA is especially useful in the resolution of mixtures of populations with differing characteristics.

The procedure of Cavalli-Sforza et al. (1994) was used with non-transformed allele frequencies at all loci. Figure 2 is based upon a Euclidean distance matrix of these loci. The total number of alleles at the 4 loci was 99, representing 95 degrees of freedom, yielding large discriminating power. The three coordinate axes in Figure 2 are 77.3% of the total genetic variation calculated as the cumulative sum of positive eigenvalues. The first principal component comprised 52% of the total variance among groups and split the populations sampled into *O. mossambicus* and *O. niloticus* groups.

As a result, we conclude that the genetic signature for *O. niloticus* is present in southern California in culture at farm #2 (by legal permit from California DFG), and in the wild (collected from the Colorado River near Blythe, CA). *O. niloticus* alleles were found at a very high level of confidence (95% level of significance) and match *O. niloticus* alleles from reference samples collected from Africa and pure lines worldwide. These fish are likely escapees from tilapia aquaculture farms in Arizona since *O. niloticus* is a legal species there. DNA markers of *O. niloticus* as a species or in hybrid forms with *O. mossambicus* were not found in any other naturalized (e.g. wild) or farmed tilapia populations sampled. Genetic confirmation that Nile tilapia (*O. niloticus*) is present near Blythe in the California region of the Colorado River implies that the species and its hybrids may be distributed in irrigation and drainage canals in the eastern part of the State. Average heterozygosities of southern California tilapias were as high as found at microsatellite loci in natural populations of other fish; however, the low heterozygosity of one farm sample (0.21) indicated the probability that a serious 'bottleneck' and/or inbreeding occurred (all other sample heterozygosities ranged 0.34–0.72) (Table 2). DNA marker data in this study also indicate that a distinct hybrid ecotype of *O. mossambicus* (likely a hybrid with *O. u. hornorum*) of unusually high heterozygosity – comparable to reference strains taken from natal populations in Africa – exists in the Salton Sea.

Significant divergence from natal African parental material has occurred in California (Figure 1, Table 2). This raises the specter that other, more isolated stocks of tilapias of high heterozygosity may occur elsewhere in southern California, and that accelerated evolutionary change is occurring. For aquaculture developers, this means that 'biological prospecting' could be as fruitful in selected, isolated locations in southern California (and elsewhere) as 'going back to Africa'. The genetic diversity of established tilapia species may be adequate to sustain and further expand an aquaculture industry into the future if planned, scientific, genetic improvement programs and investments were made.

Need to revise regulatory environment

There are worldwide concerns about importations of new, exotic species and their potential negative impacts on native aquatic biodiversity and public investments in aquatic species recovery programs. However, there is little information available to managers interested in updating regulations on the basis of more accurate and updated scientific information on biological and evolutionary changes in species since importation and establishment.

Many political entities base importation restrictions on conventional, systematic Linnean nomenclature derived from external, morphometric analyses. Anywhere in the aquatic environment where large scale introductions of exotic tilapias occur, there is a risk of interbreeding and hybridization among populations that may be morphologically and meristically distinct in Africa but still reproductively compatible due to their relatively 'recent' divergence. Wherever a mixture of tilapia species have been stocked, reproductively viable hybrids have resulted; and the use of external morphometric characterizations of these hybrids for species determinations is fruitless (Wohlfarth and Hulata 1983; Pullin 1988). The use of DNA fingerprint markers is the only viable method available to discern presence/absence of distinct tilapia species, the genetic composition of established, feral hybrids in new environments, and the composition of mixed species in culture. This study has also shown that conventional analyses are inadequate to measure the magnitude of evolutionary divergence of hybrids. Significant evolutionary divergence from African parental material was found in established tilapia populations in the wild in California. DNA marker data clearly indicate that an distinct hybrid ecotype of *O. mossambicus* (likely a hybrid with *O. u. hornorum*) of unusually high heterozygosity exists in the Salton Sea.

Species nomenclature ('Latin names') are useless for tilapiine fishes outside of the wild in Africa because the species have become so manipulated. As a result, any regulatory structure for the tilapias based upon discerning mixed hybrids is inadequate. There are likely many places like California where a certain tilapia 'species' are not permitted but the genetic material exists within that political jurisdiction in the form of hybrids. It is recommended that regulators (and aquaculture scientists) develop a system of ecotypes (code names) based upon presence of unique DNA microsatellite markers to label strains, and that hybrid strains be collected into a 'registry' of tilapias based upon DNA markers. Such a registry could be used by regulators to better manage exotic tilapias and aquaculture developments. For example, based upon this research we would propose to create two southern California ecotypes:

1. *California Mozambique (codename: alleles Os64 and Os7r):* A tilapia (a *O. u. hornorum* × *O. mossambicus* hybrid) found in the Salton Sea, in ditches in Niland, California, and used in aquaculture at two of the southern California farms sampled;
2. *California niloticus (codename: alleles Os25 and Os64):* A *O. niloticus* found in the Colorado River near Blythe, California; almost identical to a *O. niloticus* in closed system culture sampled at a southern California farm.

Acknowledgements

Dr Roger Doyle and his colleagues at the Marine Probe Laboratory of Dalhousie University, Halifax, Nova Scotia, Canada, completed the DNA and statistical analyses reported. This research was supported by the California Department of Fish and Game, Sacramento, California, under contract FG4054IF-1 to B.C.P.

References

Ambal AJD (1996) The relationship between domestication and genetic diversity of *Oreochromis* species in Malawi: *Oreochromis shiranus shiranus* (Boulenger) and *Oreochromis shiranus chilwae* (Trewavas). PhD thesis, Dalhousie University, Halifax, Nova Scotia, Canada

ATA (American Tilapia Association) (1995) 1994 Tilapia situation and outlook report. American Tilapia Association, Charlestown, West Virginia

ATA (American Tilapia Association) (2000) 1999 Tilapia situation and outlook report. American Tilapia Association, Charlestown, West Virginia

Barker J (1994) Symposium lecture. In: The Fifth World Congress on Genetics Applied to Livestock, pp 501–508. Guelph, Ontario, Canada

Barrett P (1983) Systematics of fishes of the genus *Tilapia* (Perciformes: Cichlidae) in the lower Colorado River. MS thesis, Arizona State University, Tucson

Bert T, Tringali M and Seyoum S (2002) Development and application of genetic tags for ecological aquaculture. In: Costa-Pierce B (ed) Ecological Aquaculture: the Evolution of the Blue Revolution. Blackwell Science, Oxford

Brock V (1960) The introduction of aquatic animals into Hawaiian waters. International Revue Gestamen Hydrobiologia 45: 463–480

Brodziak J, Bentley B, Bartley D, Gall G and Gomulkiewicz R (1992) Tests of genetic stock identification using coded wire tagged fish. Canadian Journal of Fisheries and Aquatic Sciences 49: 1507–1517

Brooker A, Cook D, Bentzen P, Wright J and Doyle R (1994) The organization of microsatellites differs between mammals and teleost fish. Canadian Journal of Fisheries and Aquatic Sciences 51: 1–9

Cavalli-Sforza L, Menozzi P and Piazza A (1994) The History and Geography of Human Genes. Princeton University Press, Princeton, New Jersey

Chervinski J (1971) Sexual dimorphism in *Tilapia aurea* and *Tilapia zillii* from Dor and from Lake Tiberias. Bamidgeh 23: 56–59

Chmilevskii D (1998) The influence of low temperature on the growth of *Oreochromis mossambicus*. Journal of Ichthyology 38(1): 92–99

Costa-Pierce B and Riedel R (2000) Fisheries ecology of the tilapias in subtropical lakes of the United States. In: Costa-Pierce BA and Rakocy JE (eds) Tilapia Aquaculture in the Americas, Vol 2, pp 1–20. World Aquaculture Society, Baton Rouge, Louisiana

Costa-Pierce BA, Moreau J and Pullin RSV (1993) New introductions of common carp (*Cyprinus carpio*) and their impacts on indigenous species, with reference to Sub-Saharan Africa. Discovery and Innovation, The African Academy of Sciences 5: 211–222

Courtenay W (1997) Tilapias as non-indigenous species in the Americas: environmental, regulatory and legal issues. In: Costa-Pierce BA and Rakocy JE (eds) Tilapia Aquaculture in the Americas, Volume 1, pp 18–33. World Aquaculture Society, Baton Rouge, Louisiana

DeSilva S, Perera M and Maitipe P (1984) The compsotion, nutritional status and digestibility of the diets of *Sarotherodon mossambicus* from nine man-made lakes in Sri Lanka. Environmental Biology of Fishes 11: 205–219

Engle C (1997) Marketing tilapias. In: Costa-Pierce BA and Rakocy JE (eds) Tilapia Aquaculture in the Americas, Vol 2, pp 244–257. World Aquaculture Society, Baton Rouge, Louisiana

FAO (Food and Agriculture Organization of the United Nations) (1997) Review of the state of world aquaculture. FAO Fisheries Circular. No 886, Rev 1. Inland Water Resources and Aquaculture Service, Fishery Resources Division, Rome, 163 pp

Felsenstein J (1985) Confidence limits on phylogenies: an approach using the bootstrap. Evolution 39: 783–791

Fitzsimmons K (2000) Future trends of tilapia culture in the Americas. In: Costa-Pierce BA and Rakocy JE (eds) Tilapia Aquaculture in the Americas, Vol 2, pp 252–264. World Aquaculture Society, Baton Rouge, Louisiana

Fonticiella DW and Sonesten L (2000) Tilapia aquaculture in Cuba. In: Costa-Pierce BA and Rakocy JE (eds) Tilapia Aquaculture in the Americas, Vol 2, pp 184–203. World Aquaculture Society, Baton Rouge, Louisiana

Franck, J, Harris A, Bentzen E, Denoven-Wright E and Wright J (1991) Organization and Evolution of Satellite, Minisatellite and Microsatellite DNAs in Teleost Fishes. Oxford Surveys Eukaryotic Genes, Oxford University Press, Oxford

Hauser W (1974) Biological control of aquatic vegetation in the lower Colorado River basin. Third interim report, 1974. Imperial Valley Field Station, University of California, Riverside

Hauser W (1975a) Can *Tilapia* replace herbicides? Proceedings of the Annual California-Nevada American Fisheries Society 6: 44–50

Hauser W (1975b) *Tilapia* as biological control agents for aquatic weeds and noxious aquatic insects in California. Proceedings of the Annual Conference California Mosquito Control Association 43: 51–53

Hauser W (1975c) An usually fast growth rate for *Tilapia zillii*. California Fish Game 61: 54–63

Hauser W (1977) Temperature requirements of *Tilapia zillii*. California Fish Game 63: 228–233

Herbinger C, Doyle R, Pitman E, Paquet D and Mesa K (1995) DNA fingerprint analysis of paternal and maternal effects on offspring growth and survival in communally-reared rainbow trout. Aquaculture 137: 245–256

Herbinger C, Doyle R, Taggart C, Lochmann S and Brooker A (1996) Family relationship and effective population size in a natural population of cod larvae. Canadian Journal of Fisheries and Aquatic Science

Hickling C (1960) The Malacca tilapia hybrids. Progressive Fish-Culturist 57: 1–10

Horn M (1988) The fish community of the Upper Newport Bay Ecological Reserve. In: Koerper H (ed) The Natural and Social Sciences of Orange County, pp 80–92. National Historical Foundation Orange City, Newport Beach, California

Hoover F and St Amant J (1970) Establishment of *Tilapia mossambica* Peters in Bard Valley, Imperial County, California. California Fish Game 56: 70–71

Keeley H (1957) Preliminary studies on *Tilapia mossambica* Peters relative to experimental pond culture. Proceedings of the Annual Conference of Southeast Association of Game Fish Commissioners 10: 139–149

Kimpton C, Gill P, Walton A, Urquhart A and Millican E (1993) Automated DNA profiling employing multiplex amplification of short tandem repeat loci. PCR Methods Applications 3: 13–22

Knaggs E (1976) Status of the genus *Tilapia* in California's estuarine and marine waters. Unpublished report, Operations Research Branch, California Department of Fish and Game, Chino, California, November 1976

Legner E and Fisher T (1980) Impact of *Tilapia zillii* (Gervais) on *Potamogeton pectinatus* L., *Myriophyllum spicatum* var. *exalbescens* Jepson, and mosquito reproduction in lower Colorado Desert irrigation canals. Acta Oecologia Applicata 1: 3–14

Legner E and Pelsue F (1977) Adaptations of *Tilapia* to *Culex* and chironomid midge ecosystems in South California. Proceedings of the Annual Conference California Mosquito Control Association 45: 95–97

Lobel PS (1980) Invasion by the Mozambique tilapia (*Sarotherodon mossambicus*; Pisces; Cichlidae) of a Pacific Atoll marine ecosystem. Micronesica 16: 349–355

Lombard G (1960) Four years of experimental culture of *Tilapia mossambica* in the Transvaal. Publications of the Conservation Science of Africa South of the Sahara 63: 225–228

Lopez J and Ulmer L (1983) Preliminary distribution and abundance survey of *Tilapia* found in irrigation drains in the Palo Verde Valley, Riverside County. Unpublished report, California Department of Fish and Game, Region 5, Chino, California

Lowe-McConnell R (1955) New species of tilapia from Lake Jipe and the Pangami River, East Africa. Bulletin of the British Museum of Natural History 2: 350–368

McConnel S, Hamilton L, Morris D, Cook D and Paquet D (1995) Isolation of salmonid microsatellite loci and their application to the population genetics of Canadian east coast stocks of Atlantic salmon. Aquaculture 137: 19–30

McConnell W (1965) *Tilapia*, a fish for ranch ponds. Progress in Agriculture in Arizona 17: 3–4

McConnell W (1966) Preliminary report on the Malacca *Tilapia* hybrid as a sport fish in Arizona. Progressive Fish-Culturist 28: 40–45

Millar R (1987) Maximum likelihood estimation of mixed stock fishery composition. Canadian Journal of Fisheries and Aquatic Sciences 44: 583–590

Minckley W (1973) Fishes of Arizona. Arizona Game Fish Department, Phoenix

Mironova N (1969) The biology of *Tilapia mossambica* under natural and laboratory conditions. Ichthyology 9: 506–514

Moreau J, Bambino C and Pauly D (1986) Indices of overall growth performance of 100 tilapia (Cichlidae) populations. In: Maclean J (ed) Proceedings of the First Asian Fisheries Forum, pp 201–206. Asian Fisheries Society, Manila, Philippines

Myers G (1938) Freshwater fishes and West Indian zoogeography. Smithsonian Report for 1927. Publication Number 3465: 339–364

Nei M, Tajima F and Tateno T (1983) Accuracy of estimated phylogenic trees from molecular data. Journal of Molecular Evolution 19: 153–170

Outdoor California (1964) New fish – tilapia – now under study. Outdoor California 25: 16

Pauly D, Moreau J and Prein M (1988) A comparison of overall growth performance of tilapia in open waters and aquaculture. In: The Second International Symposium on Tilapia in Aquaculture, pp 469–479. International Center for Living Aquatic Resources Management, Manila, Philippines

Pelzman R (1973) A review of the life history of *Tilapia zillii* with a reassessment of its desirability in California. California Department of Fish and Game Inland Fisheries Administrative Report 74–1, Sacramento, California

Philippart J-Cl and Ruwet J-Cl (1982) Ecology and distribution of tilapias. In: Pullin RSV and Lowe-McConnell R (eds) The Biology and Culture of Tilapias. International Center for Living Aquatic Resources Management, Manila, Philippines

Platt S and Hauser W (1978) Optimum temperature for feeding and growth of *Tilapia zillii*. Progressive Fish-Culturist 40: 105–107

Pullin R (1985) Tilapias: everyman's fish. Biologist 32: 84–88

Pullin R (1986) Culture of herbivorous tilapias. In: Chan H (ed) Proceedings of the International Conference on the Development and Management of Tropical Living Aquatic Resources, pp 145–149. University Pertanian Malaysia, Serdang, Malaysia

Pullin R (1988) Tilapia Genetic Resources for Aquaculture. International Center for Living Aquatic Resources Management, Manila, Philippines

Pullin R and Lowe-McConnell R (1982) The Biology and Culture of the Tilapias. International Center for Living Aquatic Resources Management, Manila, Philippines

Robins C, Bailey R, Bond C, Brooker J, Lachner E, Lea R and Scott W (1991) Common and Scientific Names of Fishes of the United States and Canada. American Fisheries Society Publication 20, Bethesda, Maryland

Ron B, Shimoda SK, Iwama GK and Grau EG (1995) Relationships among ration, salinity, 17alpha-methyltestosterone and growth in the euryhaline tilapia, *Oreochromis mossambicus*. Aquaculture 135: 185–193

Ruzzante D, Taggart D, Cook D and Goddard S (1996) Genetic differentiation between inshore and offshore Atlantic cod (*Gadus morhua*) off Newfoundland: microsatellite DNA variation and antifreeze level. Canadian Journal of Fisheries and Aquatic Sciences 53: 634–645

Saitu N and Nei M (1987) The neighbor joining method: a new method for reconstructing phylogenic trees. Molecular Biological Evolution 4: 406–425

Sambrook J, Fritsch E and Maniatis T (1989) Molecular Cloning: a Laboratory Manual. Cold Spring Harbor Laboratory, Cold Spring Harbor, New York

Santibanez-Koref M, Orphanos V and Boyle J (1993) Rapid determination of sequences flanking microsatellites using dephosphorylated cloning vectors. Trends in Genetics 9: 43

Schlotter C and Tautz D (1992) Slippage synthesis of simple sequence DNA. Nucleic Acids Research 20: 211–215

Schoenen P (1982) A bibliography of Important Tilapias (Pisces: Cichlidae) for Aquaculture. International Center for Living Aquatic Resources Management, Manila, Philippines

Schoenherr A (1988) A review of the life history of the desert pupfish, *Cyprinodon macularius*. Bulletin of the Southern California Academy of Sciences 87: 104–134

Schuster W (1952) Fish Culture in Brackishwater Ponds of Java. Indo-Pacific Fisheries Council Special Publication Number 1. Food and Agriculture Organization, Rome, Italy

Slatkin M (1985) Rare alleles as indicators of gene flow. Evolution 39: 53–65

St Amant J (1966) Addition of *Tilapia mossambica* Peters to the California fauna. California Fish Game 52: 54–55

Talbot F and Newell B (1957) A preliminary note on the breeding and growth of tilapia. East African Agriculture Journal 22: 118–121

Trewavas E (1973) On the cichlid fishes of the genus *Pelmatochromis* with proposal of a new genus for *P. conigus*; on the relationship between *Pelmatochromis* and *Tilapia* and the recognition of *Sarotherodon* as a distinct genus. Bulletin British Museum Natural History (Zoology) 25: 1–26

Trewavas E (1982) Generic couplings of tilapiini used in aquaculture. Aquaculture 27: 79–81

Trewavas E (1983) Tilapiine Fishes of the Genera *Sarotherodon, Oreochromis* and *Danakilia*. British Museum Natural History, London

Watanabe WO, Ellingson LJ, Wicklund RI and Olla BL (1988) The effectts of salinity on growth, food consumption and conversion on juvenile, monosex male Florida red tilapia. In: Pullin RSV, Bhukaswan T, Tonguthai K and Maclean JL (eds) The Second International Symposium on Tilapia in Aquaculture, ICLARM Conference Proceedings 15, pp 515–523. Department of Fisheries, Bangkok, Thailand/International Center for Living Aquatic Resources Management, Manila, Philippines

Whitfield A and Blaber S (1979) The distribution of the freshwater cichlid *Sarotherodon mossambicus* in estuarine systems. Environmental Biology of Fishes 4: 77–81

Wohlfarth G and Hulata G (1983) Applied Genetics of Tilapias. ICLARM Studies and Reviews 6. International Center for Living Aquatic Resources Management, Manila, Philippines

Wright J and Bentzen P (1994) Microsatellites: genetic markers for the future. Reviews in Fish Biology 4: 384–388

Ziegle J, Su Y, Corcoran Y, Nie L and Mayrand P (1992) Application of automated DNA sizing technology for genotyping microsatellite loci. Genome 14: 1026–1031

Biological Invasions **5:** 85–102, 2003.

Estuarine and scalar patterns of invasion in the soft-bottom benthic communities of the San Francisco Estuary

Henry Lee II[1,*], Bruce Thompson[2] & Sarah Lowe[2]
[1]*US Environmental Protection Agency, ORD, NHEERL, Western Ecology Division, 2111 S.E. Marine Science Drive, Newport, OR 97365, USA;* [2]*San Francisco Estuary Institute, 7770 Pardee Lane, Oakland, CA 94621-1424, USA;* **Author for correspondence (e-mail: lee.henry@epa.gov; fax: 541 867 4049)*

Received 4 October 2001; accepted in revised form 9 July 2002

Key words: estuarine gradients, invasion metrics, invasion patterns, nonindigenous species, San Francisco Estuary, soft-bottom communities, spatial scales

Abstract

The spatial patterns of nonindigenous species in seven subtidal soft-bottom communities in the San Francisco Estuary were quantified. Sixty nonindigenous species were found out of the 533 taxa enumerated (11%). Patterns of invasion across the communities were evaluated using a suite of invasion metrics based on the abundance or species richness of nonindigenous species. Patterns of invasion along the estuarine gradient varied with the invasion metric used, and the ecological interpretation of the metrics is discussed. Overall, the estuarine transition community located in the estuarine turbidity maximum zone (mean 5 practical salinity unit (psu)), main estuarine community (mean 16 psu), and marine muddy community (mean 28 psu) were more invaded than two fresh-brackish communities (mean <1 psu) and a marine sandy community (mean 27 psu). Nonindigenous species were numerically dominant over much of the Estuary, making up more than 90% of the individuals in two communities. The percentage of the total species composed of nonindigenous species increased at smaller spatial scales: 11% at the estuary (gamma) scale, 21% at the community (alpha) scale, and 42% at the grab (point) scale. Wider spatial distributions of nonindigenous species and a relatively greater percentage of rare native species may have resulted in this pattern. Because of this scale dependency, comparisons among sites need to be made at the same spatial scale. Native species were positively correlated with nonindigenous species in several of the communities, presumably due to similar responses to small-scale differences in habitat quality. The rate of invasion into the soft-bottom communities of the San Francisco Estuary appears to have increased over the last one to two decades and many of the new introductions have become numerically dominant.

Introduction

Increasingly, nonindigenous species represent a major environmental threat to nearly every aquatic and terrestrial ecosystem (e.g. Westbrooks 1998; Mack et al. 2000; Pimentel et al. 2000). Among coastal ecosystems, the San Francisco Bay-Delta (the 'San Francisco Estuary') has become the archetype for an invaded estuary. Cohen and Carlton (1998) reported 234 nonindigenous species from the Bay area, substantially more than reported from other North American bays and estuaries (Ruiz et al. 1997, 2000). Even with this plethora of introduced species, the rate of new invasions has increased from one new species every 55 weeks prior to 1960 to one every 14 weeks (Cohen and Carlton 1998). Many of these invaders have been successful and have become the numerical or biomass dominants in a number of habitats (e.g. Nichols and Thompson 1985a; Nichols and Pamatmat 1988; Carlton et al. 1990; Meng et al. 1994; Cohen and Carlton 1995). Taken together, these observations suggest that the San Francisco Estuary 'may be the most invaded estuary in the world' (Cohen and Carlton 1998).

Previous studies have documented the presence of nonindigenous species from a diversity of habitat types across the entire San Francisco Estuary (e.g. Cohen and Carlton 1995) or quantified the extent of invasion at particular sites (e.g. Nichols and Thompson 1985a). What has been lacking, however, is a quantitative evaluation of the nature and extent of invasion in a specific habitat type across the entire estuarine gradient. A recent compilation of four monitoring surveys has generated an assessment of the subtidal soft-bottom communities throughout the San Francisco Estuary (Thompson et al. 2000), from the freshwater Delta region to the marine regions of the Central Bay (Figure 1). Classification analysis on these data identified seven benthic sub-assemblages, or community types, based on species composition and abundances (Thompson et al. 2000). Results from these combined surveys and the classification of the benthos into biotic assemblages provides a data set and an ecological framework to evaluate the patterns of invasion both within and across benthic community types.

The primary objective of this study is to use the data from the four monitoring programs to quantify patterns of invasion both within and across the subtidal soft-bottom benthic communities of the San Francisco Estuary. A second objective is to evaluate how the patterns of invasion vary with different invasion metrics and the ecological interpretation of these metrics at different spatial scales. The final objective is to determine the associations between native and nonindigenous species, and to use these statistical correlations to make inferences about the processes controlling nonindigenous species.

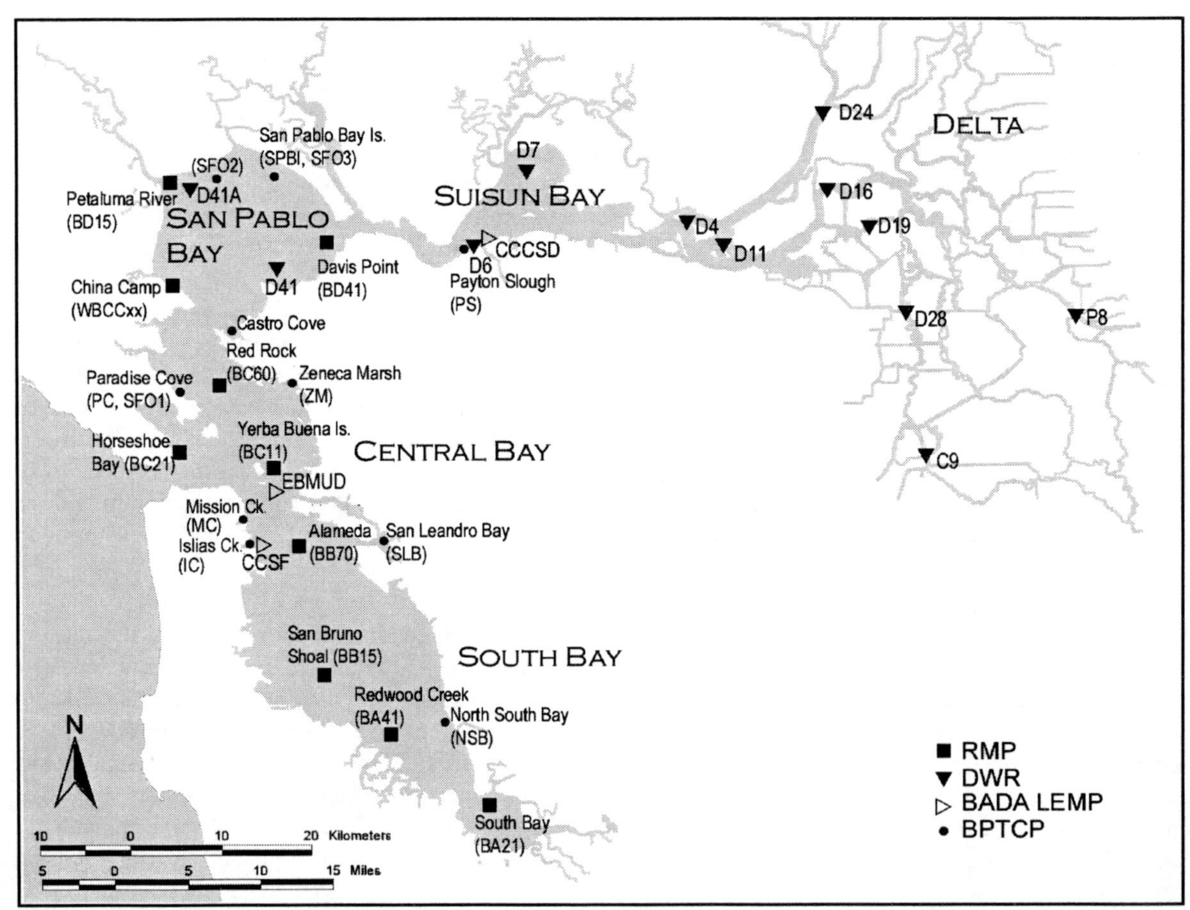

Figure 1. Location of the sampling stations in the four monitoring programs in the San Francisco Bay-Delta. RMP = Regional Monitoring Program. DWR = Department of Water Resources. BADA LEMP = Bay Dischargers Association Local Effects Monitoring Program. BPTCP = Bay Protection and Toxic Cleanup Program. The designations at the sites are the station identifiers used by the respective programs.

Methods

The present analysis combines results from four monitoring studies of the subtidal soft-bottom benthos in the San Francisco Estuary. The Regional Monitoring Program for Trace Substances (RMP) sampled nine fixed sites throughout the Bay during both the wet and dry periods from 1994 to 1997 (Thompson et al. 2000). The Bay Area Dischargers Association's Local Effects Monitoring Program (BADA LEMP) sampled nine sites within 50–350 m of sewage discharges on the same schedule as the RMP program (Thompson et al. 1999). The California Department of Water Resources (DWR) conducted monthly sampling in San Pablo Bay, Suisun Bay, and the Delta (unpublished) of which a total of 15 sites from 1994 to 1997 were included in the present analysis. Finally, the State's Bay Protection and Toxic Cleanup Program (BPTCP) sampled four sites in Castro Cove in 1992 (Carey et al. 1994; Hunt et al. 1998). These studies covered most of the Estuary, from the mouth of the Bay up into the Delta (Figure 1).

A single benthic grab was usually taken at each site in the RMP and BADA LEMP programs, three to four grabs at the DWR sites, and five grabs at the BPTCP Castro Cove site. Most samples were collected with a 0.05 m^2 Ponar grab. The samples in the BADA LEMP program collected with a 0.1 m^2 Smith-McIntyre grab were split in half and only one-half analyzed. The BPTCP samples were collected using a 0.018 m^2 grab; three of these smaller grabs were summed to approximate the area of the Ponar grabs. All studies used a 0.5 mm mesh sieve. Taxonomic nomenclature among studies was standardized, to the extent possible, by careful review of the taxon lists.

Using the 1992–1996 data from these studies, Thompson et al. (2000) used ordination and classification procedures (Smith et al. 1985) to identify benthic assemblages, or sets of samples with similar species compositions and abundances. The ordination–classification analysis was based on 424 samples collected at 44 sites and included a total of 354 species. The Bray–Curtis dissimilarity index (Bray and Curtis 1957) was used as the index of similarity among all pairs of samples. Ordination of the samples was based on principal coordinates analysis (Gower 1966, 1967), while the flexible clustering procedure (Lance and Williams 1966) was used to produce dendrograms that classified the samples and taxa based on ecological distance. Sets of samples with major faunal differences were classified as assemblages. The assemblages were further divided into sub-assemblages, which had partially overlapping species compositions but differed in composition and abundances of the dominant species. The sub-assemblages are considered equivalent to 'communities' in the present analysis. To increase the number of samples within each sub-assemblage, each of the 1997 grab samples from the RMP, BADA LEMP, and DWR studies was assigned to a benthic sub-assemblage by comparing the faunal composition of the grab to the average composition of each sub-assemblage using the Sorensen similarity index (Sorensen 1948). Adding the 1997 data increased the total number of replicates to 590 samples.

Cohen and Carlton's (1995) report and the 1999 addendum on nonindigenous species in the San Francisco Estuary were used to classify species as nonindigenous or cryptogenic, the species of uncertain origin (Carlton 1996). Additionally, we define the 'indeterminate' taxa as those taxa not identified to a sufficiently low taxonomic level (usually species) to classify as native, nonindigenous, or cryptogenic. Unless otherwise noted, 'native' refers to all species or individuals that were not specifically classified as nonindigenous, and hence include the cryptogenic species and indeterminate taxa.

Ten invasion metrics based on the absolute or relative abundance or species richness of nonindigenous species were used in assessing the extent of invasion within and across the sub-assemblages (Table 1). Numerical abundance of nonindigenous species within sub-assemblages was measured both as the mean abundance of nonindigenous species per grab (MAGnis) and as their mean relative abundance (M%AGnis). Numerically dominant species were defined as the five most abundant species within each sub-assemblage (Table 2).

Species richness of nonindigenous species was evaluated at three spatial scales. The largest was over the entire San Francisco Estuary, the gamma diversity of a landscape (i.e. across multiple communities). The second was at the scale of an individual sub-assemblage, the alpha diversity of a 'homogeneous' community. The third spatial scale was the richness within individual grabs, the point diversity within a sample from a community. The invasion metrics at the Estuary scale were the number of nonindigenous species found within the San Francisco Estuary (SEnis) and their percentage of the total number of species (%SEnis). The metrics at the community scale were the absolute number of nonindigenous species per sub-assemblage

Table 1. Invasion metrics and the parameters used in the statistical correlations or in deriving the invasion metrics.

Species richness – estuary scale
SEnis = number of nonindigenous species within the San Francisco Estuary
%SEnis = percent of total species composed of nonindigenous species
Species richness – community scale
SCt_i = total number of species within sub-assemblage i
$SCnat_i$ = number of native species within sub-assemblage i
$SCnis_i$ = number of nonindigenous species within sub-assemblage i
$\%SCnis_i$ = percent of total species within sub-assemblage i composed of nonindigenous species
OM%SCnis = overall mean percent of the species within sub-assemblages composed of nonindigenous species = (mean of SCnis in seven sub-assemblages)/(mean of SCt in seven sub-assemblages) ∗ 100
Species richness – grab scale
SGt_{ij} = total number of species within grab j within sub-assemblage i
$SGnat_{ij}$ = number of native species in grab j within sub-assemblage i
$SGnis_{ij}$ = number of nonindigenous species in grab j within sub-assemblage i
$\%SGnis_{ij}$ = percent of species in grab j in subassemblage i composed of nonindigenous species
$MSGt_i$ = mean number of total species per grab within sub-assemblage i
$MSGnat_i$ = mean number of native species per grab within sub-assemblage i
$MSGnis_i$ = mean number of nonindigenous species per grab within sub-assemblage i = average of SGnis within a sub-assemblage
$M\%SGnis_i$ = mean percent of species per grab composed of nonindigenous species within sub-assemblage i = average of %SGnis within a sub-assemblage
OM%SGnis = overall mean percent of nonindigenous species per grab = (mean of SGnis for all 590 samples)/(mean of SGt for all 590 samples) ∗ 100
Abundance
AGt_{ij} = total abundance in grab j within sub-assemblage i
$AGnat_{ij}$ = abundance of native species in grab j within sub-assemblage i
$AGnis_{ij}$ = abundance of nonindigenous species within grab j within sub-assemblage i
$\%AGnis_{ij}$ = relative abundance of nonindigenous species in grab j within sub-assemblage i
MAGt = mean abundance of all species per grab in sub-assemblage i
$MAGnat_i$ = mean abundance of native species per grab within sub-assemblage i
$MAGnis_i$ = mean abundance of nonindigenous species per grab within sub-assemblage i = average of AGnis for a sub-assemblage
$M\%AGnis_i$ = mean relative abundance of nonindigenous species per grab within sub-assemblage i = average of %AGnis for a sub-assemblage

Species richness of nonindigenous species was evaluated at the estuary scale (gamma diversity), community scale (alpha diversity), and grab scale (point diversity). 'Total species' includes all the nonindigenous and native species, including the cryptogenic species and indeterminate taxa. The 10 invasion metrics used in the analysis are bolded.

(SCnis) and the percent of the total species composed of nonindigenous species within each sub-assemblage (%SCnis). Because of their smaller sample sizes, the total number of nonindigenous species in the fresh-brackish sandy sub-assemblage ($n = 32$) and especially the estuarine margin sub-assemblage ($n = 14$) and marine sandy sub-assemblage ($n = 8$) is likely to be an underestimate. The overall mean percent nonindigenous species at the community scale (OM%SCnis) was calculated by dividing the average number of nonindigenous species in each of the seven sub-assemblages by the average of the total number species in the seven sub-assemblages. Point diversity within each of the sub-assemblages was measured both as the mean number of nonindigenous species per grab (MSGnis) and as the mean percent of the species per grab (M%SGnis). An overall mean of the percent nonindigenous species at the point scale (OM%SGnis) was calculated by dividing the average number of nonindigenous species per grab based on all 590 samples by the average total number of species per grab in the 590 samples.

The mean number of nonindigenous species per grab (MSGnis) was tested for equality among the seven sub-assemblages with the non-parametric Kruskal–Wallis test on ranks. The Kruskal–Wallis test was also used to test for equality among the sub-assemblages in the mean percent of the species per grab composed of nonindigenous species (M%SGnis). Dunn's method for multiple comparisons was used to test for differences among individual sub-assemblages. A significance value of $P = 0.05$ was used in these comparisons. These tests and all statistics were run using SigmaStat or SAS.

Table 2. Environmental characteristics and abundances and classification of the numerically dominant species in the seven benthic sub-assemblages (1992–1996 data).

Environmental characteristic/ species	Taxon	Class	Fresh-brackish sandy (FBs)	Fresh-brackish muddy (FBm)	Fresh-brackish estuarine transition (FBt)	Main estuarine (ESm)	Estuarine margin (ESi)	Marine sandy (MAs)	Marine muddy (MAm)
Mean salinity (psu)	—	—	0.08	0.7	4.9	16.1	22.8	26.6	27.5
Mean percent fines	—	—	15	72	51	88	92	5	74
Mean percent TOC	—	—	0.7	3.9	2.1	2.6	2	0.4	1
Corbicula fluminea	Pe	NIS	**19**	**50**	1				
Paratendipes sp. A	Ch	NAT	**6**	1					
*Gammarus daiberi**	A	NIS	**4**	**36**	**4**	1			
Corophium stimpsoni	A	NAT	**2**	**57**	**4**	1			
*Marenzelleria viridis**	P	NIS	**2**	6	**20**	2			
Limnodrilus hoffmeisteri	O	Crypt.	1	**36**	2				
*Potamocorbula amurensis**	Pe	NIS	1	5	**28**	**162**	3	1	4
Manayunkia speciosa	P	NIS		**117**					
Corophium alienense	A	NIS		1	**26**	**5**	16		1
Ampelisca abdita	A	NIS			1	**135**	55		**697**
*Nippoleucon hinumensis**	Cu	NIS		1	3	**22**	**145**		8
*Corophium heteroceratum**	A	NIS			1	**9**			**133**
Tubificidae	O	Indet.			1	3	**404**	1	9
Streblospio benedicti	P	NIS			1	3	**116**		1
Corophium spp	A	Indet.			1		**92**		**53**
Gemma gemma	Pe	NIS				1	**70**		1
Nematoda	N	Indet.				1	37	**8**	25
Corophium acherusicum	A	NIS				1	6	1	**745**
Euchone limnicola	P	Crypt.						1	**58**
Leptochelia dubia	T	Crypt.							50
Heteropodarke heteromorpha	P	NAT						**18**	
Hesionura coineaus difficillis	P	NAT						**3**	
Glycera tenuis	P	NAT						**2**	

The five numerical dominants in each sub-assemblage are bolded (only top four in marine sandy sub-assemblage because of ties). Abundances are per 0.05 m^2. NAT – native species; NIS – nonindigenous species; Crypt. – cryptogenic species; Indet. – indeterminate taxa. Species introduced in the 1980s or 1990s are indicated by *. A – amphipod; Ch – chironomid; Cu – cumacean; N – nematode; O – oligochaete; P – polychaete; Pe – pelecypod; T – tanaid. (Modified from Thompson et al. 2000.)

Table 3. Percent contribution of nonindigenous species to total species richness at three spatial scales within the San Francisco Estuary.

Spatial scale	Diversity type	Scale measured (abbreviation)	Mean (range)
Estuary	Gamma	San Francisco Estuary (%SEnis)	11% (NA)
Community	Alpha	Sub-assemblage (OM%SCnis)	18% (9–32%)
Individual grab	Point	Individual grabs (OM%SGnis)	36% (11–71%)

See Table 1 and the text for how the values were calculated. Ranges for the community and grab scales are the ranges of the means (%SCnis and M%SGnis) across the seven sub-assemblages.

The percent of the species composed of nonindigenous species at each of the three spatial scales (Table 3) was compared to the next smaller scale. The single value for the San Francisco Estuary (%SEnis) was compared to the community scale by comparing it to the 95% confidence interval of the mean percentage for the seven sub-assemblages (OM%SCnis, $n = 7$). The community and point scales were compared using a paired t-test that compared the difference between the percentage of nonindigenous species within each sub-assemblage (%SCnis) and the corresponding mean percent per grab (M%SGnis) for the same sub-assemblage ($n = 7$). The differences in the percents were transformed using an arcsin square root transformation.

Spearman's rank correlations were used to determine the relationship between the abundances of native and nonindigenous species per grab (AGnat and AGnis) and between the number of native and nonindigenous species per grab (SGnat and SGnis) across all seven sub-assemblages ($n = 590$). Spearman's rank correlations between native and nonindigenous species were also determined within each of the seven sub-assemblages. This resulted in 7 correlations for abundance and 7 for species richness, with between 8 and 250 replicates per sub-assemblage (Table 4). To account for the multiple comparisons, a modified Bonferroni procedure (Jaccard and Wan 1996) was used to adjust the overall alpha level to 0.05. With the modified Bonferroni procedure, any individual correlation with an alpha less than 0.0125 is considered significant.

The relationship between native and nonindigenous species richness was also evaluated at the community scale. One approach was to correlate the number of native species within each sub-assemblage (SCnat) with the corresponding number of nonindigenous species (SCnis) using Spearman's rank correlation. This correlation was limited to the four sub-assemblages with >80 samples. A second approach was to correlate the mean numbers of native and nonindigenous species per grab for each of the seven sub-assemblages (MSGnat and MSGnis, $n = 7$). In a final analysis, the relationship between the species richness of nonindigenous species at the point and community scales was examined by regressing the mean number of nonindigenous species per grab (MSGnis) on the number of nonindigenous species within the corresponding sub-assemblage (SCnis). These regressions were determined using all seven sub-assemblages and with just the four sub-assemblages with >80 samples.

Table 4. Spearman's rank correlation coefficients between native and nonindigenous species within the San Francisco Estuary and within the seven sub-assemblages.

Correlation	*N*	No. taxa	Abundances
San Francisco Estuary	590	0.111* (0.007)	0.332* (<0.001)
Fresh-brackish sandy (FBs)	32	−0.123 (0.499)	0.288 (0.109)
Fresh-brackish muddy (FBm)	250	−0.036 (0.571)	0.473* (<0.001)
Estuarine transition (FBt)	92	0.310* (0.003)	0.391* (<0.001)
Main estuarine (ESm)	113	−0.065 (0.493)	0.236* (0.012)
Estuarine margin (ESi)	14	0.466 (0.0873)	0.169 (0.552)
Marine sandy (MAs)	8	−0.407 (0.290)	−0.652 (0.072)
Marine muddy (MAm)	81	0.384* (<0.001)	0.465* (<0.001)

'No. taxa' – correlation between the numbers of native and nonindigenous species per grab ($SGnat_{ij}$ and $SGnis_{ij}$). 'Abundances' – correlation between the abundances of native and nonindigenous species per grab ($AGnat_{ij}$ and $AGnis_{ij}$). *N* – number of grab samples. Probability values are given in parentheses. The overall alpha for the suite of correlations with taxa or with abundances across the seven sub-assemblages was set at 0.05. Using the modified Bonferroni method, individual correlations were significant at an alpha of 0.0125. *Statistically significant.

Results

Benthic assemblages

Thompson et al. (2000) identified three major subtidal assemblages within the San Francisco Estuary based on their biotic composition: (1) fresh-brackish, (2) estuarine, and (3) marine. In turn, each assemblage was composed of two or three sub-assemblages differing to some degree in species composition or dominance (Table 2). In the present analysis, these biotic sub-assemblages are considered as equivalent to 'communities'. Mean salinity within the sub-assemblages ranged from less than 1 practical salinity unit (psu) to 28 psu, while sediment type varied from sandy to muddy (Table 2). The fresh-brackish assemblage occupied sites in the Delta and was composed of the fresh-brackish muddy sub-assemblage (FBm), fresh-brackish sandy sub-assemblage (FBs), and estuarine transition sub-assemblage (FBt). The estuarine transition sub-assemblage occurred within the estuarine turbidity maximum zone at the confluence of the Sacramento River, Suisun, and San Pablo Bays. The estuarine assemblage was composed of the main estuarine sub-assemblage (ESm) and the estuarine margin sub-assemblage (ESi). The main estuarine sub-assemblage was composed of sites that were spatially separated, and included stations in Suisun and San Pablo Bays as well as in the lower end of South Bay. The estuarine margin sub-assemblage was limited to the China Camp tidal channels and Castro Cove and was dominated by species often associated with disturbance such as *Streblospio benedicti* and tubificid worms. The marine assemblage occurred in the Central Bay and the mid- and upper portions of South Bay and was characterized by nearly coastal salinities except during high river inflow. Most of the marine assemblage was composed of the marine muddy sub-assemblage (MAm) with a marine sandy sub-assemblage (MAs) limited to one site (Red Rock) in the Central Bay.

Although this study incorporates samples taken over several seasons and years, the approach of assigning samples to sub-assemblages by biotic composition rather than spatial location reduces the effect of temporal variability. The species composition and abundances of the sub-assemblages (cluster groupings) were relatively stable both spatially and temporally. Additionally, most of the sub-assemblages included samples from all seasons and times sampled. If temporal variability had been the major driver, the samples would have grouped according to season and/or year. When the species composition or abundances did substantially change at a particular site, the samples were assigned to a different cluster grouping based on the changes in biotic composition. For example, most samples from the Central Bay were classified in the marine muddy sub-assemblage (Thompson et al. 2000). However, following the flood flows which reduced salinity within the Bay, samples collected from the Central Bay in the winter of 1997 were classified as part of the main estuarine sub-assemblage. Samples from those same sites were classified with the marine muddy sub-assemblage again the following summer after a return to normal salinities. Therefore, the biotic composition and community structure values for the sub-assemblages used in the analyses reflect the average conditions for the various biotic assemblages.

Abundance of nonindigenous species

The abundances of the numerically dominant species for each of the sub-assemblages are given in Table 2. Of the 23 numerically dominant species in all seven sub-assemblages, 12 were nonindigenous, 3 were cryptogenic, 5 were native, and 3 were indeterminate taxa. The introduced amphipod *Corophium acherusicum* had the greatest density of any species, averaging 14,900 m^2 in the marine muddy sub-assemblage. Mean abundance of nonindigenous species (MAGnis) varied more than 500-fold among the seven sub-assemblages and about 40-fold if the marine sandy sub-assemblage was excluded (Figure 2). These differences among sub-assemblages were highly significant as tested by the Kruskal–Wallis test ($P < 0.001$). The multiple comparisons test identified the marine muddy, estuarine margin, and main estuarine sub-assemblages as the group with the greatest abundance of nonindigenous

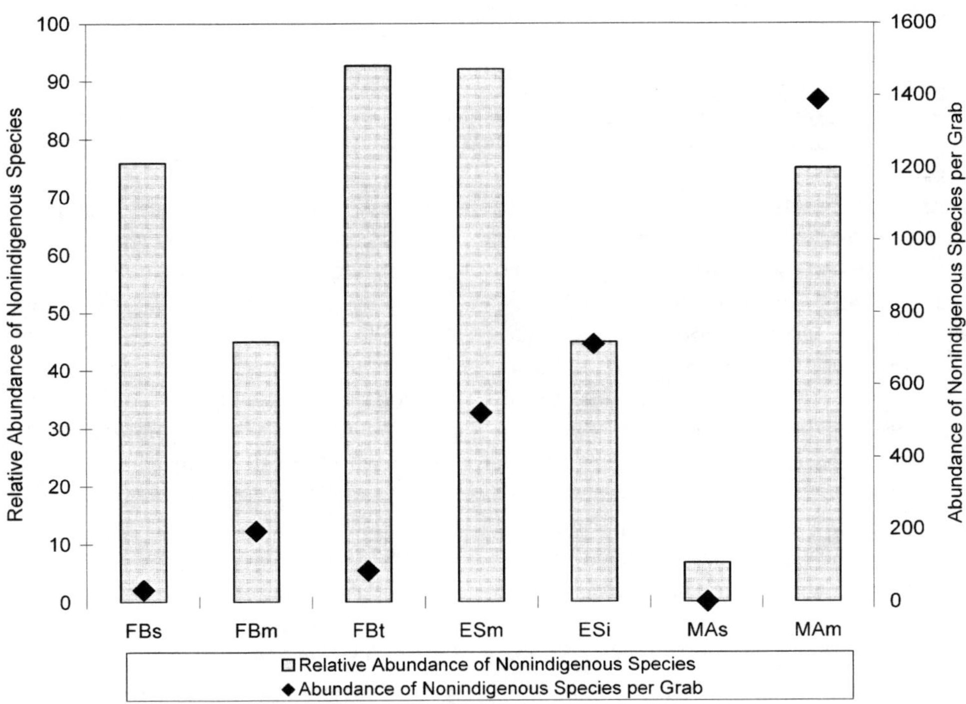

Figure 2. Mean abundance of nonindigenous species per 0.05 m^2 grab (MAGnis) and mean relative abundance of nonindigenous species per grab (M%AGnis) in the subtidal benthic sub-assemblages in the San Francisco Estuary. Sub-assemblages ordered from lowest to highest mean salinity. FBs – fresh-brackish sandy sub-assemblage. FBm – fresh-brackish muddy sub-assemblage. FBt – estuarine transition sub-assemblage. ESm – main estuarine sub-assemblage. ESi – estuarine margin sub-assemblage. MAm – marine muddy sub-assemblage. MAs – marine sandy sub-assemblage.

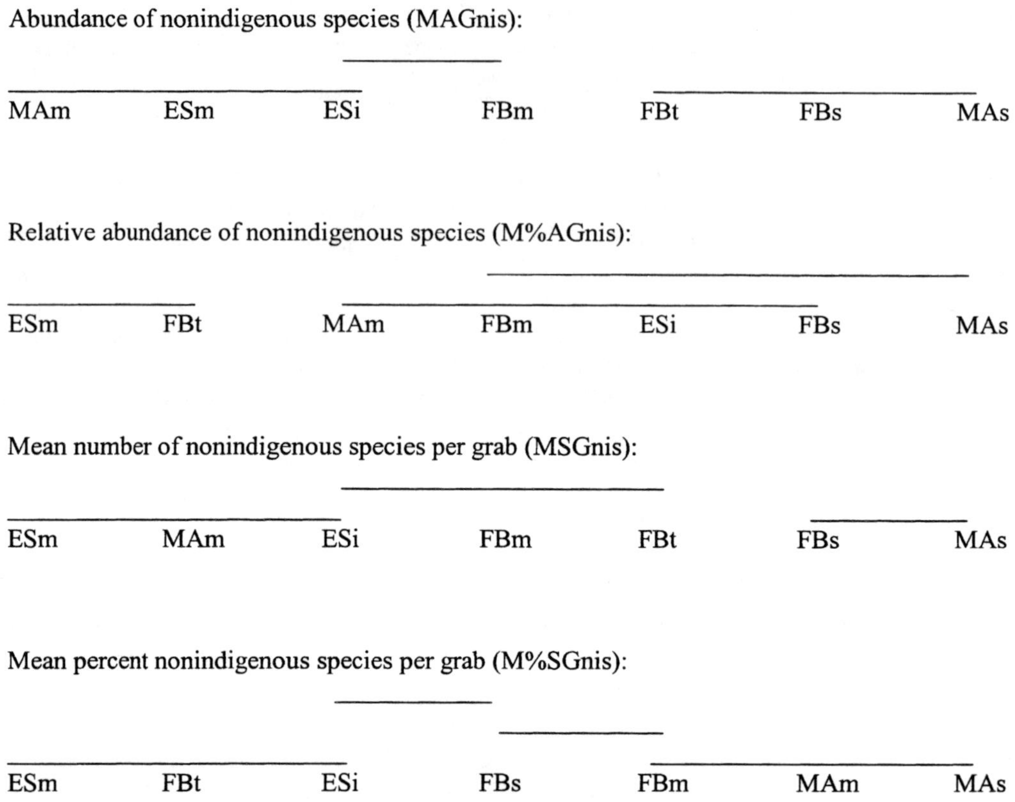

Abundance of nonindigenous species (MAGnis):

MAm ESm ESi FBm FBt FBs MAs

Relative abundance of nonindigenous species (M%AGnis):

ESm FBt MAm FBm ESi FBs MAs

Mean number of nonindigenous species per grab (MSGnis):

ESm MAm ESi FBm FBt FBs MAs

Mean percent nonindigenous species per grab (M%SGnis):

ESm FBt ESi FBs FBm MAm MAs

Figure 3. Results of multiple comparisons tests for abundance and species richness of nonindigenous taxa across the seven sub-assemblages. Sites not connected by a line are significantly different (Dunn's' method, $P < 0.05$). FBs – fresh-brackish sandy sub-assemblage. FBm – fresh-brackish muddy sub-assemblage. FBt – estuarine transition sub-assemblage. ESm – main estuarine sub-assemblage. ESi – estuarine margin sub-assemblage. MAm – marine muddy sub-assemblage. MAs – marine sandy sub-assemblage.

species (Figure 3). The multiple comparisons test also identified the marine sandy, freshwater sandy, and estuarine transition sub-assemblages as the group with the lowest abundance of nonindigenous species. Mean relative abundance of the nonindigenous species in the seven sub-assemblages (M%AGnis) ranged from 7% to 93%. Inclusion of the cryptogenic species increased the range to 24–99% and had the greatest impact on the marine sandy sub-assemblage, the site with the lowest relative abundance. The differences in relative abundance of nonindigenous species among sub-assemblages were highly significant as tested by the Kruskal–Wallis test ($P < 0.001$). The multiple comparison test grouped the estuarine muddy and estuarine sub-assemblages as the community types with the highest relative abundances of nonindigenous species.

Species richness of nonindigenous species

A total of 60 nonindigenous benthic species were identified within the San Francisco Estuary (SEnis), constituting 11% of the 533 taxa identified (%SEnis). Inclusion of the 17 cryptogenic species increased the percentage to 14%. Exclusion of the indeterminate taxa increased the percentage of nonindigenous species to 19% while inclusion of the cryptogenic species and exclusion of the indeterminate taxa increased the value to 24%. As previously noted, the present analysis includes the cryptogenic species and indeterminate taxa with the native species in all calculations, therefore the 60 nonindigenous species and the 11% values represent the measures of gamma diversity of nonindigenous species. Of these 60 nonindigenous species, polychaetes, amphipods, and bivalves were the most diverse groups, comprising 22%, 18%, and 15% of the introduced species, respectively.

At the community scale, nonindigenous species constituted 9–32% of the species within the seven sub-assemblages (%SCnis) (Figure 4), while inclusion of the cryptogenic species increased these percentages to 19–41%. The number of nonindigenous species per sub-assemblage (SCnis) in the four sub-assemblages

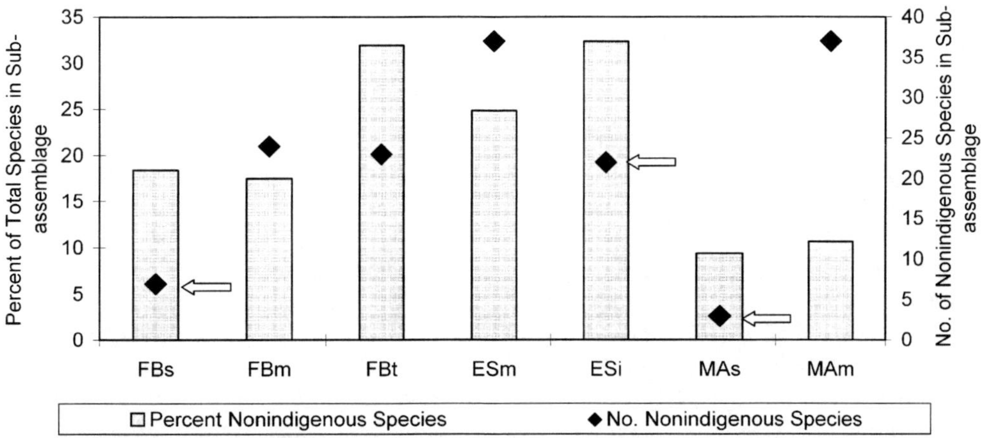

Figure 4. Number of nonindigenous species per sub-assemblage (SCnis) and percent of total number of species per sub-assemblage composed of nonindigenous species (%SCnis) in the subtidal benthic sub-assemblages in the San Francisco Estuary. The absolute numbers of nonindigenous species need to be interpreted cautiously in the fresh-brackish sandy ($n = 32$), estuarine margin ($n = 14$), and marine sandy ($n = 8$) sub-assemblages because of the smaller number of samples (identified by arrows). Sub-assemblages ordered from lowest to highest mean salinity. FBs – fresh-brackish sandy sub-assemblage. FBm – fresh-brackish muddy sub-assemblage. FBt – estuarine transition sub-assemblage. ESm – main estuarine sub-assemblage. ESi – estuarine margin sub-assemblage. MAm – marine muddy sub-assemblage. MAs – marine sandy sub-assemblage.

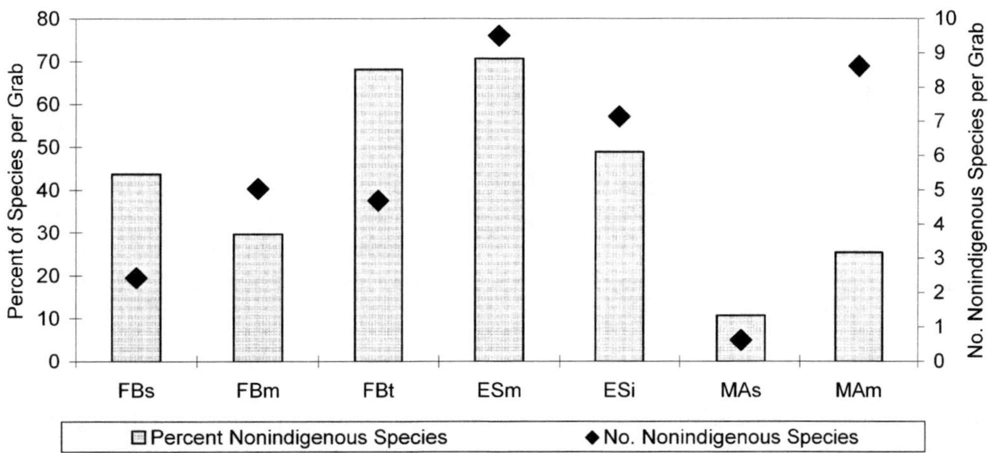

Figure 5. Mean number of nonindigenous species per grab (MSGnis) and mean percentage of total species per grab composed of nonindigenous species (M%SGnis) in the subtidal benthic sub-assemblages in the San Francisco Estuary. Sub-assemblages ordered from the lowest to the highest mean salinity. FBs – fresh-brackish sandy sub-assemblage. FBm – fresh-brackish muddy sub-assemblage. FBt – estuarine transition sub-assemblage. ESm – main estuarine sub-assemblage. ESi – estuarine margin sub-assemblage. MAm – marine muddy sub-assemblage. MAs – marine sandy sub-assemblage.

with >80 samples ranged from 23 to 37 species (Figure 4). At the point scale, the average number of nonindigenous species per grab (MSGnis) ranged from less than 1 to almost 10 (Figure 5), while inclusion of the cryptogenic species increased the range from 1 to 12. The differences in the mean number of nonindigenous species per grab among the sub-assemblages were highly significant as tested by the Kruskal–Wallis test ($P < 0.001$). The multiple comparison test separated the main estuarine, marine muddy, and estuarine margin sub-assemblages as the group with the greatest number of nonindigenous species per grab (Figure 3). On a relative basis, nonindigenous species comprised an average of 11–71% of the species per grab (M%SGnis, Figure 5), while inclusion of the cryptogenic species increased the range to 23–73%. The differences in the percentages of nonindigenous species per grab among the sub-assemblages were highly significant as tested by the Kruskal–Wallis test ($P < 0.001$). The multiple comparisons test separated the main estuarine, estuarine transition, and estuarine margin sub-assemblages as the group with the highest percentage composition of nonindigenous species per grab (Figure 3).

The percentage contribution of nonindigenous species to species richness increased at smaller spatial scales (Table 3). The 11% value at the estuary scale (%SEnis) was about 60% of the overall mean for the community scale (OM%SCnis) and fell outside the 95% confidence interval for the mean percent invasion for the seven sub-assemblages (%SCnis, 95% confidence interval of 12–29%). In turn, the 18% overall mean at the community scale (OM%SCnis) was half the overall mean at the grab scale (OM%SGnis). This difference between the point and community scales was statistically significant as tested using a paired t-test comparing the percentage of nonindigenous within each sub-assemblage (%SCnis) with the mean percentage for grabs (M%SGnis) within the same sub-assemblages ($P < 0.05, n = 7$).

Statistical associations

There was a weak but significant positive rank correlation between the numbers of native and nonindigenous species per grab (SGnat and SGnis) and a stronger positive correlation between the abundances of native and nonindigenous species (AGnat and AGnis) using all 590 samples (Table 4). Within the sub-assemblages, 6 of the 14 rank correlations between the abundances or species richnesses of native and nonindigenous species were significantly positive (Table 4). Four of these positive correlations were between the abundances of nonindigenous and native species per grab (AGnat and AGnis) and two were between the number of native and nonindigenous species per grab (SGnat and SGnis). In contrast to the correlations within sub-assemblages, the numbers of native and nonindigenous species per sub-assemblage (SCnat and SCnis) were not significantly correlated (rho$=0.63$, $P= 0.33$). Nor were the mean abundances of native and nonindigenous species per sub-assemblage (MSGnat and MSGnis) correlated (rho$=0.21$, $P = 0.60$). There was a significant linear relationship between the mean number of nonindigenous species per grab (MSGnis) and the number of nonindigenous species within the corresponding sub-assemblage (SCnis) whether tested using all seven sub-assemblages ($R^2 = 0.91$, $P < 0.01$) or just the four sub-assemblages with >80 samples ($R^2 = 0.98$, $P < 0.05$) (Figure 6).

Discussion

Combining the four monitoring programs in the San Francisco Estuary resulted in a data set of almost 600 samples covering seven subtidal benthic sub-assemblages from freshwater to essentially marine conditions. As such, this data set allows an examination of the nature and extent of invasion within a variety of soft-bottom community types and how the pattern of invasion changes along an estuarine gradient.

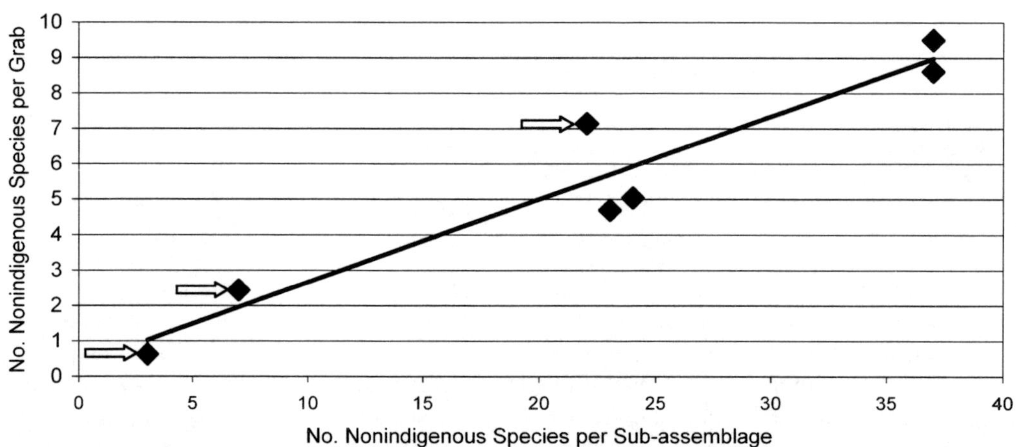

Figure 6. Relationship between the point diversity (MSGnis) and alpha diversity (SCnis) of nonindigenous species. The relationship was significant whether based on all seven sub-assemblages ($R^2 = 0.91$, $P < 0.01$) or just the four sub-assemblages with >80 samples ($R^2 = 0.98$, $P < 0.05$). The arrows identify the three sub-assemblages with <80 samples. Increasing the number of samples would likely increase the number of nonindigenous species within these sub-assemblage and have the effect of shifting the point to the right.

By having data at the grab, community, and Estuary scales, it is also possible to evaluate the relationship between spatial scale and the extent of invasion. In making these comparisons, we utilize a suite of invasion metrics, or indicators, (Table 1) both to capture the different types of effects nonindigenous species have on benthic community structure and to evaluate the utility of the invasion metrics.

Patterns of nonindigenous species abundance

Nonindigenous species were abundant in six of the subtidal benthic communities of the San Francisco Estuary, with mean densities (MAGnis) ranging from about 800–28,000 m^2 (Figure 2). The marine sandy sub-assemblage was the only exception, which as discussed below may be considered a harsh environment. These subtidal abundances are lower than reported for a mudflat in the South Bay, which frequently had densities of nonindigenous species exceeding 100,000 m^2 (Nichols and Thompson 1985a). While the mean subtidal densities were lower, patches within the marine muddy and main estuarine sub-assemblages equaled the intertidal densities, with a maximum nonindigenous density of 341,000 m^2 in one sample. With the exception of the marine sandy sub-assemblage, the highest abundances of nonindigenous species occurred in the higher salinity sub-assemblages (Figures 2 and 3). However, it seems that the salinity zones of highest nonindigenous species abundance varies with specific estuarine conditions. In two Oregon estuaries, maximum nonindigenous densities occurred at salinities as low as 5 psu and as high as 25 psu depending upon tidal height and the estuary (Castillo 2000; also see Castillo et al. 2000).

Relative abundance of nonindigenous species (M%AGnis) ranged from 7% to 93% (Figure 2), with the oligohaline estuarine transition and mesohaline main estuarine sub-assemblages having the highest relative percentages. With more than 90% of the individuals consisting of introduced species, these two communities are essentially alien assemblages. The Asian clam, *Potamocorbula amurensis,* was the most abundant species in both sub-assemblages (Table 2). However, even if *Potamocorbula* was excluded (from both MAGt and MAGnis), nonindigenous species still constituted about 90% of the remaining individuals, with *Marenzelleria viridis* and *Corophium alienense* as the numerical dominants in the estuarine transition sub-assemblage and *Ampelisca abdita* as the dominant in the main estuarine sub-assemblage (Table 2). Similarly, nonindigenous species still constituted about 25–60% of the remaining individuals in the freshwater muddy, freshwater sandy, and estuarine margin sub-assemblages after the exclusion of the single most abundant nonindigenous species from the calculations. These results reflect the 'depth' of invasion in these communities. These communities are invaded by so many nonindigenous species that a population crash of the numerically dominant introduced species would not greatly alter the relative abundance of nonindigenous species, though it would affect absolute abundance.

Patterns of nonindigenous species richness

The species richness of nonindigenous species was evaluated at three spatial scales (see Whittaker 1977; Magurran 1988; Scott and Jennings 1998): (1) gamma diversity at the scale of the San Francisco Estuary, (2) alpha diversity at the scale of the sub-assemblages or communities, and (3) point diversity at the scale of the individual grabs (Table 3). At the Estuary scale, gamma diversity was equal to the 60 nonindigenous species identified from all the sub-assemblages (SEnis) or, on a relative basis, 11% of the species (%SEnis). Perhaps surprisingly given San Francisco Estuary's reputation as the most invaded estuary, introduced species comprised only a moderate percentage of the total soft-bottom fauna.

Wolff's (1999) analysis of the macrobenthos in the brackish (5–20 psu) and high salinity (>20 psu to seawater) waters in the Netherlands is one of the few studies where it is possible to compare both the number and percent of nonindigenous species at an 'estuary' scale. Even including a few species that would not have been collected in the present study (e.g. *Teredo*), the Dutch estuaries contained less than half the number of nonindigenous species compared to San Francisco: 16 in the brackish waters, 14 in the higher salinities, and a total of 27 in both salinity zones. From these data it is possible to estimate that the total estuarine fauna in the Netherlands consist of between 8% and 11% nonindigenous species, depending upon the extent of overlap between the brackish and high salinity faunal lists. It appears that other estuaries may have similar percentages of nonindigenous fauna as the San Francisco Estuary even though their absolute numbers of nonindigenous species may be considerably less.

Maximum alpha diversity, as measured by the number of nonindigenous species within a sub-assemblage (SCnis), was 37 introduced species in both the main estuarine and marine muddy sub-assemblages (Figure 4). Mean salinity was just below the limit for polyhaline zones in the main estuarine sub-assemblage and at the upper limit for polyhaline zones for the marine muddy sub-assemblage (Table 2). In comparison, the number of introduced species was about a third less in the oligohaline fresh-brackish muddy and mesohaline estuarine transition sub-assemblages. This pattern of a greater diversity of introduced species at higher salinities was also observed when comparing the fresh-brackish sandy and estuarine margin sub-assemblages even though more than twice the number of samples were taken in the freshwater community (32 vs. 14 grabs). The one exception to this pattern was the marine sandy sub-assemblage, which likely reflects both the small sample size and the harshness of this environment. The pattern of a greater species richness of nonindigenous species at higher salinities is consistent with the conclusion by Ruiz et al. (2000) that there are more invasions at higher salinities. In contrast, Wolff (1999) reported similar numbers of introduced species in brackish (5–20 psu) and high salinity waters (>20 psu) in the Netherlands, though with his salinity classes it is not possible to determine whether the diversity of nonindigenous species declines at a salinity of about 5 psu as observed in the San Francisco Estuary.

Based on their relative contribution to the total fauna within communities (%SCnis), the highest percentage invasion occurred between 5 and 23 psu (Figure 4). The decline in the percentage in the marine muddy sub-assemblage reflects the high native diversity in this community. The similarity in the percent invasion in the marine muddy and marine sandy sub-assemblages (11% vs. 9%) demonstrates that communities with very different numbers and densities of nonindigenous species can have a similar relative species richness of introduced species. Accordingly, a low relative percent invasion in a species rich community needs to be interpreted cautiously.

Maximum point diversity, as measured by the mean number of nonindigenous species per grab (MSGnis), occurred over a mean salinity range of 16–28 psu (Figure 5, Table 3). Again, the marine sandy sub-assemblage was the exception. This pattern in point diversity appeared to be related to the alpha diversity of the community as the number of nonindigenous species per grab was significantly related to the number of nonindigenous species within the community (Figure 6). Though perhaps not totally unexpected, such a relationship would not occur if the nonindigenous species occupied specialized microhabitats within a community, obscuring any relationship between point and alpha diversities. In any case, about 20–30% of the nonindigenous species within a community were found within the 0.05 m^2 grabs. This suggests that as additional species invade a community, the point diversity of nonindigenous species will increase proportionately, at least within certain bounds. In terms of their relative contribution to point diversity (M%SGnis), nonindigenous species constituted a substantial if not major proportion of the fauna within grabs (25–71%) except in the marine sandy sub-assemblage (Figure 5). The major difference with the pattern for relative *versus* absolute species richness was a shifting of the zone of maximum invasion to a lower salinity range, 5–23 psu.

Comparison of the extent of invasion across communities

The availability of data covering a range of benthic assemblages raises the question as to which were the most and least invaded communities. The marine sandy sub-assemblage was the least invaded by every measure. The low percentage of fine-grained sediment (Table 2) suggests that this site is subjected to strong currents and sediment instability. The low organic content of the sediment may also limit the benthic community. Either in isolation or in combination (see Snelgrove and Butman 1994), these abiotic factors may create a 'harsh' environment, limiting colonization to a small number of relatively specialized species. The low total abundance (MAGt) and total species per grab (MSGt) are consistent with a harsh environment, while the limitation of three of the four numerically dominant species to this site (Table 2) is consistent with a community inhabited by 'specialized' species. Other factors, such as propagule pressure (number of individuals introduced in an invasion event and number of invasion events, e.g. Forsyth and Duncan 2001; Kolar and Lodge 2001) or an inherently greater biological resistance to invasion, may also have played a role.

Identifying the most highly invaded communities is more problematic since the ranking depends upon the invasion metric. One would come to very different conclusions about the extent of invasion in the marine muddy sub-assemblage depending upon whether the absolute or relative species richness per grab (MSGnis or M%SGnis) was used (Figure 5). Furthermore, the pattern for the abundance of nonindigenous species (MAGnis) did not resemble any of the other metrics (Figure 2). Even with this variation among metrics, it is possible to discern some general patterns. The estuarine margin, main estuarine, and marine muddy sub-assemblages were the most highly invaded communities based on the absolute measures of nonindigenous species richness (SCnis and MSGnis) as well as their numerical abundance (MAGnis). These communities covered a salinity range of 16–28 psu and were composed of fine-grained sediments. Based on the relative metrics (%SCnis, M%SGnis, and M%AGnis), the most invaded communities shifted to the estuarine transition and main estuarine sub-assemblages. These communities had mean salinities ranging from 5 to 16 psu. The estuarine margin community also was highly invaded based on the percentage of the species within the community (%SCnis) but moderately invaded based on the relative measures for species per grab (M%SGnis) and abundance (M%AGnis). The lowest salinity sites, the fresh-brackish sandy and muddy sub-assemblages, displayed a moderate level of invasion by most metrics but were not ranked among the most invaded by any of them.

This analysis assumes that the estuarine gradient in salinity was the dominant controlling factor, which is supported by the ordination analysis on these benthic data identifying salinity as the most highly correlated variable with species composition and abundance (Thompson et al. 2000). Additionally, salinity has been hypothesized as one of the major factors affecting the within estuary (Ruiz et al. 2000) and biogeographic patterns of marine/estuarine invasions (Chapman 2000). However, other abiotic and biotic factors play a role in structuring benthic communities, and presumably contributed to determining the invasion patterns in the San Francisco Estuary. In particular, sediment characteristics and the associated hydrological processes are likely to affect the diversity and abundance of introduced species. As mentioned, the number of nonindigenous species in the marine sandy sub-assemblage is thought to be limited by sediment instability. The effects of sediment characteristics are also suggested by the lower abundance and species richness per grab of nonindigenous species (MAGnis and MSGnis, respectively) in the fresh-brackish sandy community compared to the fresh-brackish muddy community even though salinities were similar.

Effect of spatial scale on the percentage species richness of nonindigenous species

One pattern arising from the present analysis was an increase in the percentage contribution of nonindigenous species to species richness at smaller spatial scales (Table 3). The percentage of nonindigenous species increased almost two-fold from the Estuary scale (%SEnis, gamma diversity) to the community scale (OM%SCnis, alpha diversity) and by another two-fold to the grab scale (OM%SGnis, point diversity). We hypothesize this pattern resulted from the nonindigenous species containing a greater proportion of generalist species with wider within-habitat and among-habitat distributions than the native species. Consequently, as the spatial scale increased, and different habitats and microhabitats were sampled, new native species would have accumulated more rapidly than the wide-spread nonindigenous species. Wide environmental ranges and phenotypic plasticity are often stated characteristics of 'weedy' invaders (Sakai et al. 2001; but see Kolar and Lodge 2001) and is consistent with the observation that nearly all marine invaders occupy more than one salinity zone (Ruiz et al. 2000). For the San Francisco benthos, the relative distributions of nonindigenous and native species can be evaluated by comparing the within-habitat and among-habitat distributions of the numerically dominant species. Within the sub-assemblages where they were a numerical dominant (Table 2), the nonindigenous species occurred in an average of 73% of the grabs *versus* 58% for the native species (frequencies calculated from Thompson et al. 2000), consistent with broader within-habitat distributions. Nonindigenous species also had wider among-habitat distributions as indicated by 11 of the 12 numerically dominant nonindigenous species occurring in 3 or more sub-assemblages in comparison to only 1 of the 5 numerically dominant native species.

Differences in the species-abundance distributions may have been an additional mechanism. To the extent that native species contained a greater percentage of rare species, relatively more new native species would have been accumulated as the spatial scale, and number of samples, increased. Regardless of the exact mechanism, these results demonstrate the importance of matching spatial scales when comparing the extent of invasion among sites. In particular, care should be taken not to compare the percent invasion within an estuary (gamma scale) and a single community (alpha scale).

Table 5. Interpretation and utility of invasion metrics based on the abundance or species richness of nonindigenous soft-bottom benthic species.

Metric	Ecological interpretation/utility
Abundance	1. Rank relative risks among communities
	2. Rank relative risks from different NIS
A: Abundance of NIS (MAGnis)	3. Strength of potential NIS-native interactions
	4. Resource utilization by NIS
R: NIS as % of total abundance (M%AGnis)	5. Use to monitor population control efforts
	R-1. Comparable among sites with different abundances
Gamma diversity (estuary scale)	1. Scale of local extinctions
	2. NIS pool available for invasion within estuary
	3. NIS pool available for invasion of other estuaries
A: No. NIS within estuary (SEnis)	4. Measure of potential risk readily understandable to public and managers
R: NIS as % of total species within estuary (%SEnis)	*A-1. Use to monitor effectiveness of vector control (e.g. ballast water treatment)*
	A-2. Difficult to compare among sites or over time if based on qualitative approaches
	R-1. Comparable among sites with different species richnesses or sampling intensities
Alpha diversity (community scale)	1. Indicator of the susceptibility to invasion and/or propagule pressure
	2. Rank relative risks among communities
A: No. NIS within community (SCnis)	*A-1. Species richness frequently used indicator of community condition*
R: NIS as % of total species within community (%SCnis)	*A-2. Difficult to compare among sites or over time if based on qualitative approaches*
	R-1. Comparable among sites with different species richnesses
Point diversity	1. Indicator of number of NIS-native biotic interactions at scale of individual benthic organisms
A: No. NIS per grab (MSGnis)	2. Readily compared among sites and over times if similar sampling methods used
R: NIS as % of total species within grab (M%SGnis)	3. Relationship to extent of invasion at community and estuary scales not clear
	R-1. Comparable among sites with different species richnesses

'A' – metrics related to absolute measures of abundance or species richness. 'R' – metrics related to relative measures of abundance or species richness. Factors only pertaining to absolute or relative metrics are numbered and italicized. NIS – nonindigenous species.

Comparison of invasion metrics

The use of a suite of invasion metrics raises the questions of how to interpret the various metrics and which is the 'best' to use in monitoring and/or control programs. As a guide to the metrics, we summarize their ecological interpretation and utility as indicators of ecological condition (Table 5). Metrics based on numerical abundance of nonindigenous species are the most appropriate indicators of the strength of biotic interactions with native species and of resource utilization. In terms of diversity, the absolute numbers of nonindigenous species within an estuary (SEnis) or community (SCnis) are equivalent to the most commonly used meaning of 'diversity' and are readily understandable by the public and managers. Also, these absolute metrics relate to a number of key invasion processes such as invasibility. Both of these absolute invasion metrics can be derived from quantitative surveys, as in the present study, or qualitatively from compilations (e.g. Cohen and Carlton 1995) or rapid assessment surveys (e.g. Cohen et al. 1998). While the qualitative approaches have proven critical in defining the scope of the ecological threat posed by invasive species within geographic areas, their results are difficult to compare among sites or over time. However, the qualitative surveys are much less costly than quantitative surveys. Relative measures of the species richness of nonindigenous species (%SEnis and %SCnis) allow comparisons among estuaries or communities with inherently different species richnesses. However, this can result in similar estimates of invasion in communities with greatly different numbers of nonindigenous species, as demonstrated by the comparison of the marine sandy and muddy sub-assemblages (Figure 4).

One advantage of the point diversity measures (MSGnis and M%SGnis) is that they do not require estimates of the total number of native and nonindigenous species within a community or estuary. As such, they are the most readily compared among sites or over time and are amenable to standard statistical approaches (e.g. ANOVA), as long as comparable sampling techniques are used. Grabs approximate the area over which benthic species interact, and as such are probably the best metrics to estimate the effects of invasion on the 'neighborhood' scale (e.g. Levine 2000). What is not clear, however, is how metrics of point diversity relate to invasion at the community or estuary scales, which are often the scales of greater interest to managers and the public. While point diversity was related to alpha diversity within the San Francisco Estuary (Figure 6), it is not known whether this relationship holds in other estuaries. If they are not related, it would be possible to have a constant measure of invasion as measured by the number of introduced species per grab while the number of invaders within the community or estuary increased.

Table 5 is an initial attempt at characterizing a suite of invasion metrics based on benthic community structure for use in ecological monitoring and research. Such a characterization is only an initial step and these invasion metrics, or any other invasion metrics, need to be rigorously evaluated for their suitability for monitoring programs and management decisions (Jackson et al. 2000). We also recognize that metrics are indicators, not reality. As true for most diversity indices, the invasion metrics used here inherently assume all species have equal effects on other species and on ecological functions as well as having equal social value. This is often not the case, and these metrics should not be used without considering the actual species composition of the native and nonindigenous species. Finally, it is also important to recognize that these metrics are subject to various types of errors or biases, especially when comparing among sites. Adequacy of the taxonomy is a critical example. Other examples of biases are discussed in Ruiz et al. (2000).

Associations between native and nonindigenous species

A basic tenet of invasion biology is that more diverse communities are less susceptible to invasion (e.g. Elton 1958; MacArthur 1972; Case 1990), though several studies have found a positive relationship between species diversity and the number of introduced species (Levine and D'Antonio 1999). In the present study, there was no significant correlation between the number of nonindigenous and native species across the communities when tested using the numbers per sub-assemblage (SCnat and SCnis) or the mean numbers per grab (MSGnat and MSGnis). Although the sample size at the community scale is limited (n = 4 or 7), these results failed to find any suggestion of an inverse relationship between native species richness and the number of invaders.

The previous correlations evaluated the relationship between invasibility and diversity at the community level. A different question is whether nonindigenous and native species are correlated within individual grab samples. Because of the numerical dominance by nonindigenous species, we expected to find negative correlations between nonindigenous and native species. However, when examined over the entire San Francisco Estuary, there was a positive correlation between the abundances of native and nonindigenous species (Table 4). There also was a significant but weaker correlation between the species richness of native and nonindigenous species over the San Francisco Estuary. Positive correlations were also found within sub-assemblages, where 6 of the 14 correlations were significant (Table 4). The most likely cause for these positive associations is that native and nonindigenous species respond in a similar fashion to within-habitat differences in habitat quality, such as small-scale differences in the percent fine-grain sediments. Localized disturbances, such as predation (e.g. Levin 1984), also could promote positive associations. This proposed mechanism is consistent with Levine and D'Antonio's (1999) conclusion that introduced and native species respond to the same environmental factors. Temporal synchronicity of the native and nonindigenous species does not appear to have been a major factor in generating these positive correlations as the classification analysis did not group the samples by month or year.

Facilitation between nonindigenous and native species, as has been found with zebra mussels (Bially and MacIsaac 2000), is an alternative explanation. For example, the introduced clam *Potamocorbula* exposes the end of its shell above the sediment potentially creating habitat for colonization by fouling organisms (Carlton et al. 1990). However, obligate fouling organisms constituted only a minor percentage of the species and were not numerically dominant in any of the sub-assemblages. It is possible that there are more subtle positive interactions between native and nonindigenous species (e.g. dense beds *Potamocorbula* reducing predation). However, positive correlations were found in four sub-assemblages ranging from freshwater to near marine conditions. To invoke facilitation as a general mechanism would require positive biotic interactions among different suites of nonindigenous and native species, which seems unlikely.

Rates of invasion

Cohen and Carlton (1998) reported an accelerating rate of invasion for the San Francisco Estuary based on all the nonindigenous fauna and flora from all habitats. An increased invasion rate is also apparent within the soft-bottom communities. Seven of the fifty-eight soft-bottom nonindigenous species from this study were introduced in the 1980s and another five by the mid-1990s (dates of introduction from Cohen and Carlton 1995; the nonindigenous *Macoma petalum* and *Mytilus galloprovincialis* excluded because of uncertain dates of introduction). This is equivalent to an invasion rate of 0.67 species/year from 1980 through 1997. Taking 1851 as a starting point for invasions into the San Francisco Estuary (Cohen and Carlton 1998), an invasion rate of 0.36 species/year would account for the 46 species introduced from 1851 through 1979. Ballast water is a likely vector of introduction for 10 of the 12 introductions over the last two decades (based on modes of introduction in Cohen and Carlton 1995 and Ruiz et al. 2000), coinciding with the period over which the total tonnage of cargo coming into San Francisco Estuary ports increased (Cohen 1998).

These recent invaders have been disproportionately successful as evidenced by constituting almost half (5 of 12) of the numerically dominant nonindigenous species (Table 2). In particular, the estuarine transition and main estuarine sub-assemblages have been highly altered by the Asian clam *Potamocorbula amurensis,* which was first recorded in 1986 (Carlton et al. 1990). The high densities of this filter-feeding clam are associated with reduced recruitment of the previously dominant 'dry-period' benthic community (Nichols et al. 1990), seasonal phytoplankton blooms (Alpine and Cloern 1992), and copepod populations (Kimmerer et al. 1994). The rapidity of the resulting changes in benthic community structure appears to be a recent phenomenon as 17 years ago Nichols and Thompson (1985a) concluded that there had been no 'significant long-term changes in the species composition of the benthos in any of the major embayments over the last three decades' (also see Nichols and Thompson 1985b).

The success of these recent invaders over the last two decades may simply represent the chance invasion of particularly invasive species. Alternatively, recent climatic events may have created conditions favorable for invaders. The importance of salinity fluctuations is suggested by the association of the establishment of *Potamocorbula* with the flood of early 1986, which resulted in nearly depauperate benthic conditions (Nichols et al. 1990), and the invasion of introduced fish in Suisun Marsh with the drought of 1986–1992 (Meng et al. 1994). The adaptability of the recent benthic invaders to flood events is suggested by all five of the numerically dominant recent invaders extending into the fresh-brackish portion of the Estuary (Table 2). Since 90% of the freshwater input into the San Francisco Estuary occurs through the Sacramento and San Joaquin Rivers (Walters et al. 1985), Suisun and San Pablo Bays are more susceptible to rapid and extreme environmental fluctuations from either floods or droughts. Accordingly, it is possible to speculate that new invasions are more likely to become established initially in Suisun and San Pablo Bays and then migrate to the Central and South Bays, as was found with *Potamocorbula* (Carlton et al. 1990).

Disclaimer

This information has been funded wholly (or in part) by the US Environmental Protection Agency. It has been subjected to the Agency's peer and administrative review, and it has been approved for publication as an EPA document. Mention of trade names or commercial products does not constitute endorsement or recommendation for use.

Acknowledgements

This study was possible only through access to the benthic data from the Department of Water Resources (DWR), BADA LEMP, and BPTCP programs. The authors thank Rusty Fairey (Calif. Dept. Fish and Game, Moss Landing, BPTCP), Jim Salerno (BADA), Heather Peterson and Cindy Messer (DWR) for making the data available and helpful discussions. Taxonomists on the studies included Jim Oakden and Peter Slattery (Moss Landing Marine Laboratory), Wayne Fields (Hydrozoology, Sacramento), and Michael Kellogg, Kathy Langan-Cranford, Patricia McGregor, and Brain Sak (City and County of San Francisco). The classification and ordination analysis was conducted by Dr Bob Smith and Ms Laura Riege. The original analysis and classification of the monitoring data were partially supported by BADA to Bruce Thompson. US EPA Region 9 funded a workshop at the San Francisco Estuary Institute on the analysis of these data in terms of nonindigenous species. We thank Dr Winona Victery and Luisa Valiela (US EPA Region 9) for their assistance in supporting this workshop. Henry Lee thanks Dr David Klauder and Carl Kohnert for their support while he was the ORD Regional Scientist in US EPA Region 9. Dr Bruce Boese, Dr John Chapman, Dr Judith Pederson, Debbie Reusser, and David Cassell provided helpful reviews on various drafts of the paper. The analysis and presentation of the data were substantially enhanced by comments from four anonymous reviewers.

References

Alpine AE and Cloern JE (1992) Trophic interactions and direct physical effects control phytoplankton biomass and production in an estuary. Limnology and Oceanography 37: 946–955

Bially A and MacIsaac HJ (2000) Fouling mussels (*Dreissena* spp.) colonize soft sediments in Lake Erie and facilitate benthic invertebrates. Freshwater Biology 43: 85–97

Bray J and Curtis J (1957) An ordination of the upland forest communities of southern Wisconsin. Ecological Monographs 27: 325–349

Carlton JT (1996) Biological invasions and cryptogenic species. Ecology 77: 1653–1654

Carlton JT, Thompson J, Schemel L and Nichols FH (1990) Remarkable invasion of San Francisco Bay (California, USA) by the Asian clam *Potamocorbula amurensis*. I. Introduction and dispersal. Marine Ecology Progress Series 66: 81–94

Carey D, Oliver J and Iampietro P (1994) Amphipod toxicity and benthic community analysis. In: San Francisco Estuary Pilot Regional Monitoring Program: Sediment Studies, Final Report for the San Francisco Regional Water Quality Control Board and the State Water Resources Control Board, Sacramento, California

Case TJ (1990) Invasion resistence arises in strongly interacting species-rich model competition communities. Proceedings of the National Academy of Sciences, USA 887: 9610–9614

Castillo GC (2000) Benthic biological invasions in two temperate estuaries and their effects on trophic relations of native fish and community stability. Doctoral dissertation, Oregon State University, 251 pp

Castillo GC, Li HW and Rossignol PA (2000) Absence of overall feedback in a benthic estuarine community: a system potentially buffered from impacts of biological invasions. Estuaries 23: 275–291

Chapman J (2000) Climate effects on the geography of nonindigenous peracaridan crustacean introductions in estuaries. In: Pederson J (ed) Proceedings of the First International Conference on Marine Bioinvasions, pp 66–80. MIT, Cambridge, Massachusetts

Cohen AN (1998) Ship's Ballast Water and the Introduction of Exotic Organisms into the San Francisco Estuary: Current Status of the Problem and Options for Management. San Francisco Estuary Institute, Oakland, California, 81 pp

Cohen AN and Carlton JT (1995) Nonindigenous aquatic species in a United States estuary: a case study of the biological invasions of the San Francisco Bay and Delta. Report for the National Sea Grant College Program, Connecticut and the US Fish and Wildlife Service, Washington, DC, Report no. PB 96-166525

Cohen AN and Carlton JT (1998) Accelerating invasion rate in a highly invaded estuary. Science 279: 555–558

Cohen AN, Mills CE, Berry H, Wonham MJ, Bingham B, Bookeim B, Carlton JT, Chapman JW, Cordell JR, Harris LH, Klinger T, Kohn A, Lambert CC, Lambert G, Li K, Secord D and Toft J (1998) Report of the Puget Sound Expedition, September 8–16, 1998: a rapid assessment survey of non-indigenous species in the shallow water of Puget Sound. For the Washington State Department Natural Resources, Olympia, Washington, and the US Fish and Wildlife Service, Lacey, Washington, 37 pp

Elton CS (1958) The Ecology of Invasions by Animals and Plants. Methuen, London

Forsyth DM and Duncan RP (2001) Propagule size and relative success of exotic ungulate and bird introductions to New Zealand. American Naturalist 157: 583–595

Gower J (1966) Some distance properties of latent root and vector methods used in multivariate analysis. Biometrica 53: 325–338

Gower J (1967) Multivariate analysis and multidimensional geometry. The Statistician 17: 13–28

Hunt JW, Anderson BS, Phillips BM, Newman J, Tjeerdema R, Stephenson M, Puckett M, Fairly R, Smtih RW and Taberski K (1998) Evaluation and use of sediment reference sites and toxicity tests in San Francisco Bay. Final report for California State Water Resources Control Board, Sacramento, California

Jaccard J and Wan CK (1996) LISREL Approaches to Interaction Effects in Multiple Regression. Sage Publications, Thousand Oaks, California

Jackson LE, Kurtz JC and Fisher WS (eds) (2000) Evaluation Guidelines for Ecological Indicators US EPA, EPA/620/R-99/005, 107 pp

Kimmerer WJ, Gartside E and Orsi JJ (1994) Predation by an introduced clam as the probable cause of substantial declines in zooplankton in San Francisco Bay. Marine Ecology Progress Series 113: 81–93

Kolar CS and Lodge DM (2001) Progress in invasion biology: predicting invaders. Trends in Ecology and Evolution 16: 199–204

Lance G and Williams W (1966) A generalized sorting strategy for computer classifications. Nature 212: 218–225

Levin L (1984) Life history and dispersal patterns in a dense infaunal polychaete assemblage: community structure and response to disturbance. Ecology 65: 1185–1200

Levine JM (2000) Species diversity and biological invasions: relating local process to community pattern. Science 288: 852–854

Levine JM and D'Antonio C (1999) Elton revisited: a review of evidence linking diversity and invasibility. Oikos 87: 15–26

MacArthur RH (1972) Geographical Ecology: Patterns in the Distributions of Species. Harper and Row, New York

Mack R, Simberloff D, Lonsdale WM, Evans H, Clout M and Bazzaz F (2000) Biotic invasions: causes, epidemiology, global consequences and control. Ecological Applications 10: 689–710

Magurran A (1988) Ecological Diversity and Its Measurement. Princeton University Press, Princeton, New Jersey, 179 pp

Meng L, Moyle P and Herbold B (1994) Changes in abundance and distribution of native and introduced fishes of Suisun Marsh. Transactions of the American Fisheries Society 123: 498–507

Nichols FH and Pamatmat MM (1988) The ecology of the soft-bottom benthos of San Francisco Bay: a community profile. US Fish and Wildlife Service Biological Report 85, 73 pp

Nichols FH and Thompson JK (1985a) Persistence of an introduced mudflat community in South San Francisco Bay, California. Marine Ecology Progress Series 24: 83–97

Nichols FH and Thompson JK (1985b) Time scales of change in the San Francisco Bay benthos. Hydrobiologia 120: 121–138

Nichols FH, Thompson JK and Schemel LE (1990) Remarkable invasion of San Francisco Bay (San Francisco, CA) by the Asian Clam *Potamocorbula amurensis*. II. Displacement of a former community. Marine Ecology Progress Series 66: 95–101

Pimentel D, Lach L, Zuniga R and Morrison D (2000) Environmental and economic costs associated with non-indigenous species in the United States. Bioscience 50: 53–65

Ruiz G, Carlton J, Grosholz E and Hines A (1997) Global invasions of marine and estuarine habitats by non-indigenous species: Mechanisms, extent, and consequences. American Zoologist 37: 621–632

Ruiz G, Fofonoff P, Carlton J, Wonham M and Hines A (2000) Invasion of coastal marine communities in North America: Apparent patterns, processes, and biases. Annual Review of Ecology and Systematics 31: 481–531

Sakai AK, Allendorf FW, Holt JS, Lodge DM, Molofsky J, With KA, Baughman S, Cabin RJ, Cohen JE, Ellstrand NC, McCauley DE, O'Neil P, Parker IM, Thompson JN and Weller SG (2001) The population biology of invasive species. Annual Review of Ecology and Systematics 32: 305–332

Scott JM and Jennings MD (1998) Large-area mapping of biodiversity. Annals of Missouri Botanical Garden 85: 34–47

Smith RW, Bernstein BB and Cimberg RL (1985) Community–environmental relationships in the benthos: application of multivariate analytical techniques. In: Soule DF and Kleppel GS (eds) Marine Organisms as Indicators, pp 247–326. Springer-Verlag, New York

Snelgrove PV and Butman CA (1994) Animal–sediment relationships revisited: cause *versus* effect. Oceanography and Marine Biology Annual Reviews 32: 111–117

Sorensen T (1948) A method of establishing groups of equal amplitude in plant sociology based on similarity of species count. Biologiske Skrifter 5: 1–34

Thompson B, Lowe S and Gravitz L (1999) Sediment conditions near wastewater discharges in San Francisco Bay. Report to Bay Area Discharger's Association, San Francisco Estuary Institute, Richmond, California, 40 pp

Thompson B, Lowe S and Kellogg M (2000) Results of benthic pilot study 1994–1997. Part I – macrobenthic assemblages of the San Francisco Bay-Delta, and their responses to abiotic factors. San Francisco Estuary Institute Technical report 39, 40 pp

Walters R, Cheng R and Conomos J (1985) Time scales of circulation and mixing processes of San Francisco Bay waters. Hydrobiologia 129: 13–36

Westbrooks R (1998) Invasive Plants, Changing the Landscape of America: Fact Book. Federal Interagency Committee for the Management of Noxious and Exotic Weeds (FICMNEW), Washington, DC, 109 pp

Whittaker RH (1977) Evolution of species diversity in land communities. In: Hecht MK, Steere WC and Wallace B (eds) Evolutionary Biology, Vol 10, pp 1–67. Plenum, New York

Wolff W (1999) Exotic invaders of the meso-oligohaline zone of estuaries in The Netherlands: why are there so many? Helgoländer Meeresuntersuchungen 52: 393–400

Biological Invasions **5:** 103–115, 2003.

History of aquatic invertebrate invasions in the Caspian Sea

Igor A. Grigorovich*, Thomas W. Therriault & Hugh J. MacIsaac
*Great Lakes Institute for Environmental Research, University of Windsor, Windsor, Ontario, Canada N9B 3P4; *Author for correspondence (e-mail: igrigorovich@hotmail.com; fax: +1-519-971-3616)*

Received 18 July 2001; accepted in revised form 25 April 2002

Key words: interchange, invasion, molecular clock, nonindigenous species, palaeogeography, Ponto-Caspian, stocking

Abstract

Incorporation of the fossil record and molecular markers into studies of biological invasions provides new historical perspectives on the incidence of natural and human-mediated invasions of nonindigenous species (NIS). Palaeontological, phylogeographic, and molecular evidence suggests that the natural, multiple colonizations of the Caspian basin via transient connections with the Black Sea and other basins played an important role in shaping the diversity of Caspian fauna. Geographically isolated, conspecific Ponto-Caspian lineages that currently inhabit fragmented habitats in the Ponto-Caspian region show limited genetic divergence, implying geologically recent episodes of gene flow between populations during the Pliocene to Pleistocene. Several molluscan lineages in the Caspian Sea may have descended from Lake Pannon stock before the Late Miocene isolation of the Caspian depression, about 5.8 million years ago. Anthropogenic activities during the 20th century were responsible for a 1800-fold increase in the rate of establishment of new aquatic species in the Caspian Sea compared to the preceding two million years of natural colonization. The observed success of NIS invasions during the 20th century was due primarily to human-mediated transport mechanisms, which were dominated by shipping activities (44%). Human-mediated species transfer has been strongly asymmetrical, toward the Volga Delta and Caspian Sea from or through Black and Azov Seas. Global and regional trade, particularly that mediated by commercial ships, provides dispersal opportunities for nonindigenous invertebrates, indicating that future invasions in the Caspian Sea are anticipated.

Introduction

The enclosed Caspian Sea is the world's largest brackish waterbody, comprising nearly 40% of the earth's continental surface water (Mordukhai-Boltovskoi 1960; Dumont 1998). One of the remarkable aspects of its fauna is the high level of endemism (Dumont 2000). Among the 950 extant aquatic metazoan species recorded, 424 taxa are endemic or shared partly with the Black–Azov and Aral Seas (Mordukhai-Boltovskoi 1964, 1979; Kasymov 1987). Several groups of crustaceans (e.g. Mysidacea, Cumacea, Amphipoda, and Onychopoda) and molluscs (e.g. Cardiidae and Pyrgulidae) have produced species flocks (Mordukhai-Boltovskoi 1960, 1979). At least 159 species have ranges that encompass naturally isolated habitats in the Caspian Sea, freshened areas, inlets, and estuaries of the Black and Azov Seas, and lower reaches of rivers draining into the Black, Azov, and Caspian Seas basins (hereafter, Ponto-Caspian region) (Mordukhai-Boltovskoi 1960). There is little understanding of the origin of these discontinuous distributions, despite well-known distributional patterns for some species (Mordukhai-Boltovskoi 1960; Dumont 2000; Richter et al. 2001).

Since the Late Miocene isolation of the Caspian depression, its fauna has been regarded as a 'relict' of the preceding Sarmatian Sea, which existed from 14.5 to 8.3 million years ago (mya) during the Miocene (Sovinskii 1904; Dumont 1998, 2000; Richter et al. 2001). However, fossil deposits strongly indicate morphological discontinuities among biotas of successive phases of the Caspian Sea from the Miocene through to the Pliocene (Mordukhai-Boltovskoi 1960; Zenkevich 1963). In their concept of species, Mordukhai-Boltovskoi (1960) and Zenkevich (1963) assumed that morphologically recognized species in the fossil record differed in their genetic makeup. The fossil record also indicates that the Early Pliocene marked the earliest appearance of true Caspian lineages, which subsequently disappeared for at least three million years, before reappearing in the Late Pliocene (Mordukhai-Boltovskoi 1960).

Mordukhai-Boltovskoi (1960) suggested that several Ponto-Caspian molluscan species descended from lineages that originated in Lake Pannon, a Miocene–Pliocene enclosed basin that existed in central Europe from 12 to 4 mya. Spectacular adaptive radiation in several molluscan groups produced nearly 100% endemism in Lake Pannon, which Geary et al. (2000) attributed to the lake's antiquity and physical stability (e.g. salinity of 5‰). Conversely, the Caspian basin experienced dramatic salinity fluctuations and phases of transgressions and regressions during the Tertiary and Quaternary (Mordukhai-Boltovskoi 1960; Zenkevich 1963). Several Pannonian lineages (e.g. *Dreissena*, Cardiidae) likely penetrated into the Caspian depression after the re-union of Lake Pannon with the Pontic basin (Mordukhai-Boltovskoi 1960).

The advent of genetic techniques now provides an opportunity for estimation of the relative timing of the divergence of extant lineages under the assumption of a molecular clock (e.g. Cristescu et al. 2001). Among the most useful techniques that are now employed to explore past gene flow events are allozyme and DNA sequence analyses (e.g. Berg and Garton 1994; Stepien et al. 1999; Cristescu et al. 2001; Berg et al. 2002). Cristescu et al. (2001) established that the Ponto-Caspian onychopod *Cercopagis pengoi* lineage split during the Pleistocene, 0.8 mya, forming two divergent mitochondrial (mt) DNA lineages in the Caspian and Black–Azov basins. This finding supports Mordukhai-Boltovskoi's (1960) hypothesis of geologically recent (i.e. Late Pliocene and Pleistocene) episodes of faunal migration between the Caspian and Black Sea basins.

Attempts to compare the extent of palaeobiological (i.e. Late Pliocene–Holocene) invasions with modern human-mediated introductions in the Caspian Sea have been constrained by incomplete historical information of faunal compositions (e.g. Rass 1978; Nikolaev 1979). The primary focus of this study is to appraise the accelerating rate of human-mediated invasions as compared to natural historical levels. We combine fossil, molecular, and phylogeographic evidence to estimate the extent of invasions resulting from natural mechanisms of dispersal during the last two million years (i.e. Late Pliocene–Holocene), as well as those mediated by anthropogenic activities during the 20th century. Coupling fossil records with knowledge of physical history of the Caspian depression allows us to explore dispersal vectors and resulting distributional patterns for species throughout geologic time.

Historic context of faunal turnovers in the Caspian depression

The great age of the Caspian Sea once led biologists to postulate that the modern Ponto-Caspian fauna is the remnant of the ancient Sarmatian Sea (Table 1, Figure 1A). A veliger-like, planktonic larval stage in development of Ponto-Caspian bivalves is the only known evidence that can be interpreted to reflect their marine origin (Orlova 2000). Nevertheless, the earliest appearance of Ponto-Caspian lineages in the Caspian basin is recorded in fossil deposits of the Early Pliocene (about 5 mya) age (Mordukhai-Boltovskoi 1960). Several Ponto-Caspian hydrobiid and rissoid gastropod species were recognized from older deposits of Miocene (Sarmatian and Maeotic) age (V.V. Anistratenko, Institute of Zoology, Kiev, Ukraine, pers. comm.; Figures 1A, B). However, authorities disagree with respect to the species-level classification of these morphologically cryptic molluscan groups. In most fossil groups there exist no true Ponto-Caspian taxa even at the genus level. Found in Sarmatian clay, six gammarid amphipod species, belonging to the genera *Praegmelina*, *Andrussowia*, and *Gammarus*, are morphologically divergent from modern Ponto-Caspian taxa (Derzhavin 1951; Mordukhai-Boltovskoi 1960). Furthermore, fossil molluscs from Lake Pannon exhibit greater affinities to modern Ponto-Caspian species than to fossil Sarmatian species. *Dreissena* spp. are among the lineages that originated in Lake Pannon, then spread to the Pontic Sea, and later evolved into Ponto-Caspian lineages (Geary et al. 2000). Among the 900 molluscan species described from Lake Pannon, several lineages of dreissenids, cardiids, and melanopsids have long fossil records extending back to the Miocene (Geary et al. 2000). Situated in close geographic proximity

Table 1. Summary of faunal turnovers related to geologic events in the Caspian Sea and adjacent waters based on geologic-stratigraphical studies of deposits.

Geologic epoch (mya)	Time (mya)	Link? Caspian depression	Link? Black Sea depression	Main events related to Caspian depression and its fauna
Miocene 23–5.3	14.5–12.5	Early Sarmatian (20–30‰)		Caspian, Black Sea and Pannonian depressions united in Sarmatian Sea; Tethys and Sarmatian lineages
	12.5–10.0	Middle Sarmatian (17–20‰)		Lake Pannon (5–12‰) separated from Sarmatian Sea
	10.0–8.3	Late Sarmatian (6–17‰)		freshening; extinction of marine lineages; evolution of endemic brackish-water lineages
	8.3–6.4	Maeotic (6–15‰)		Transient link with Mediterranean Sea (30–40‰); expansion and extirpation of Mediterranean lineages
	6.4–5.8	Pontic (10–15‰)*		Transgression and reconnection with Lake Pannon (5‰)*; freshening; diversification of Pontic lineages
	5.8–5.2	Babadjan (15–30‰)	Late Pontic (10–15‰)*	Basins separated; extirpation of Ponto-Caspian lineages; evolution of Babadjanian brackish-water lineages
Pliocene 5.3–1.8	5.2–3.4	Balakhan (0.1‰ or 300‰?)	Cimmerian (5–12‰)*	Extinction of Babadjanian species; expansion of freshwater species; diversification of Pontic lineages in Cimmerian Sea
	3.4–2.0	Akchagyl (5–12‰)§	Kuyalnik (5–12‰)*	Evolution of Akchagylian brackish-water lineages; major Akchagylian transgression; link with Kuyalnik Sea
	2.0–0.7	Apsheron (5–12‰)*	Gurian (5–8‰)* Chauda (5–8‰)*	Major Apsheronian transgression; transient link with Kuyalnik and Gurian Seas; expansion of Ponto-Caspian species
Pleistocene 1.8–0.01	0.7–0.35	Baku (5–10‰)*	Urundzhik (5-8‰)*	Turkyan regression during Mindel glaciation; transgression and link with Chauda Sea, faunal interchange
	0.35–0.25	Early Khazar (5–12‰)*	Ancient Euxinian (7–22‰)*	Interglacial; transgression and link with Ancient Euxinian Sea, faunal interchange; transient link between Black Sea phases and Mediterranean basin (30–40‰), intrusion of Mediterranean species
	0.25–0.06	Late Khazar (5–12‰)*	Uzunlar (5–30‰)*	Transgression; regression during Riss glaciation; salination of Uzunlar Sea, expansion of Mediterranean species
	0.06–0.019	Early Khvalyn (8–12‰)*	Karangat (30–40‰) Girkan (5–20‰)*	Major transgression and reconnection with Girkan Sea, re-colonization of Girkan basin by Ponto-Caspian species; disconnection of Girkan basin from Mediterranean Sea
	0.019–0.009	Late Khvalyn (3–13‰)*	Surozhsk (5–10‰)* New Euxininan (3–7‰)*	Transgressive-regressive stages; expansion of Ponto-Caspian species in Surozhsk and New Euxininan Seas; intrusion of Mediterranean species in Late Khvalyn Sea
Holocene 0.010–present	0.009–present	New Caspian (<1–13‰)* modern Caspian (<1–13‰)*	Ancient Black Sea (<13–22‰)*	Reconnection of Ancient Black Sea with Mediterranean Sea (30–40‰) and transition to second Mediterranean phase, colonization by Mediterranean species
			Modern Black Sea (<13–22‰)* Modern Azov Seas (<1–12‰)*	Human-mediated invasions by nonindigenous species from around world

Summarized from Derzhavin 1951; Mordukhai-Boltovskoi 1960; Zenkevich 1963; Chepalyga 1985; Geary et al. 2000.
Presence of Ponto-Caspian lineages are indicated by ∗, except for sole occurrence of *Dreissena* spp. indicated by §. Salinity is shown in parentheses. Caspian and Black Sea depressions were united until Late Miocene, about 5.8 mya. Subsequent transient reconnections between concurring basins in Caspian and Black Sea depressions are indicated with arrows.

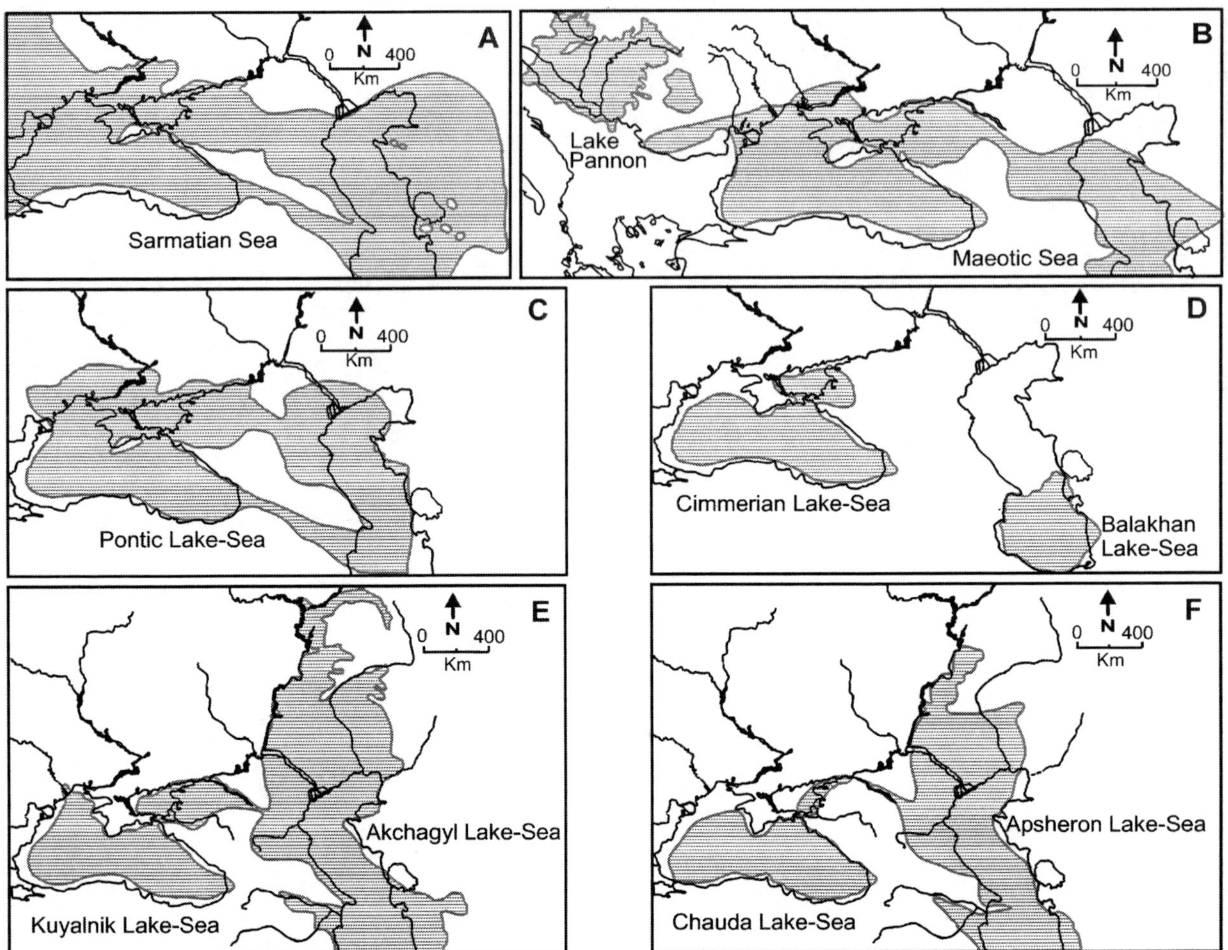

Figure 1. Map of the Ponto-Caspian region showing successive geologic basins in Caspian and Black Sea depressions (shaded) and modern boundaries of Caspian, Black and Azov Seas: Sarmatian Sea (A), Maeotic Sea (B), Pontic Lake-Sea (C), Balakhan and Cimmerian basins (D), Akchagyl and Kuyalnik (E), Apsheron and Chauda basins (F). Modified from Zenkevich (1963). See Table 1 for geologic time scale.

to the Pontic Sea depression, Lake Pannon became united with the former basin during the Late Miocene (Figures 1B, C, Table 1). The newly established seaway facilitated mass eastward migration of Pannonian fauna to the Pontic Sea (Zenkevich 1963). This major colonization pulse is marked in the fossil record of Pliocene (Pontic) age by the appearance of dreissenids and cardiids, including Ponto-Caspian lineages of *Didacna*, *Monodacna*, *Dreissena rostriformis*, and *D. polymorpha* (see Zenkevich 1963 for additional taxa).

After the Pliocene elevation of the Caucasus Mountains, the Caspian and Black Sea depressions became separated (Figures 1C, D). This isolation had a devastating impact on the Pliocene Caspian Sea biota, resulting in mass extinction of Ponto-Caspian taxa with a prior history in the basin (Mordukhai-Boltovskoi 1960; Zenkevich 1963). Detailed analyses of stratigraphical ranges of fossil invertebrates suggest that the aquatic fauna of the Caspian basin was neither ecologically nor genetically continuous during the Pliocene to Pleistocene (Table 1). Ponto-Caspian lineages had disappeared from deposits of Babadjanian, Balakhanian, and Akchagylian age (Figures 1D, E). These extinctions are not an artifact of an incomplete Pliocene–Pleistocene fossil record in the Caspian Sea, since the fossil record of the same age is rich and continuous in sediments underlying the Black Sea (Mordukhai-Boltovskoi 1960). We interpret the 3.8 million years' absence of Ponto-Caspian species in the Pliocene–Pleistocene Caspian deposits as evidence of their extirpation.

The opening of transient connections between the Caspian and Black Sea depressions during the Pliocene and Pleistocene played an important role in shaping the biotic composition of the Caspian Sea (Table 1). These connections facilitated several successive migrations of faunal components between these basins (Mordukhai-Boltovskoi 1960; Zenkevich 1963). For example, during the major Akchagylian transgression, a connection temporarily re-established between the Akchagyl and Kuyalnik Seas through the Manych depression (Figure 1E). This probably facilitated the recolonization of at least two species – *Dreissena rostriformis* and *D. polymorpha* – that once lived in the Caspian depression (Mordukhai-Boltovskoi 1960).

Beginning in the Late Pliocene (about 2.0 mya), Ponto-Caspian lineages first appeared in great numbers in the sedimentary layers and have produced a continuous fossil record that extends to the present (see Mordukhai-Boltovskoi 1960 and Zenkevich 1963 for a list of species). Many taxa previously confined to the Black Sea depression during Kuyalnik and Chaudinian phases expanded their distributions to the Apsheron Sea via the Kuma–Manych depression (Figure 1F). This flow of immigrants from western basins probably proceeded into subsequent phases of the Caspian basin including the Baku and Khazar Seas (Table 1).

Succession of glacial and interglacial cycles occurred throughout the Russian plain during the Pleistocene (about 1.6 mya to 10,000 years ago), resulting in drastic salinity fluctuations and phases of transgressions and regressions of the Caspian basin (Zenkevich 1963; Chepalyga 1985; Dumont 1998). Either during the Pleistocene or immediately following, at least 24 Arctic invertebrate taxa penetrated the Caspian basin with glacial meltwater (Orlova 2000). In the Holocene, a transient link was re-established across the Caucasus region, enabling invasion of more than 10 Atlantic–Mediterranean invertebrate species into the Caspian Sea (Rass 1978; Orlova 2000). Radiocarbon dating of *Cerastoderma* (Mollusca) shells indicates that this intrusion occurred during the Middle Holocene (Chepalyga 1985), but Derzhavin (1951) and Rass (1978) suggested that their appearance was related to an earlier link during the Late Khvalyn transgression (Table 1). Holocene Atlantic–Mediterranean invaders are represented by the bivalves *Cerastoderma lamarcki* and *C. umbonatum*, polychaete *Fabricia sabella*, bryozoan *Bowerbankia imbricata*, copepod *Calanipeda aquae-dulcis*, and several platyhelminth species (Mordukhai-Boltovskoi 1960; Kasymov 1982; Chepalyga 1985; Chepalyga and Tarasov 1997). Atlantic–Mediterranean platyhelminthes lost their native hosts and switched to native Caspian clupeid and acipenserid fishes, implying that the Atlantic–Mediterranean faunal component was likely more speciose during the Holocene than at present (Mordukhai-Boltovskoi 1960). Chepalyga and Tarasov (1997) suggested that some of these invaders were transported with ancient reed boats sailing the Kuma–Manych waterway. However, no historical evidence has been presented to confirm the existence of this entry vector.

Studies of recent sedimentary layers in the Caspian Sea revealed modern invasion pulses, which coincide with stocking efforts during the early 20th century and with the opening of the Volga–Don Canal in 1952 (Chepalyga and Tarasov 1997; Figure 2). Hence, coupling the fossil record with knowledge of physical history of the Caspian Sea provides insights into colonization events and faunal turnovers over geologic time, although precise timing of these events is rarely possible to determine using only palaeontological and geologic evidence. Consequently, critical evaluation of invasion histories, in light of molecular clock data, is essential for our understanding of biological invasions.

Pliocene–Holocene invasions

Although Ponto-Caspian lineages are known from Late Miocene Caspian deposits (see Table 1), we interpret their subsequent absence over the following 3.8 million years as true evidence of their extirpation. We therefore consider Pliocene–Holocene invaders to be those that were unknown in Babadjanian, Balakhanian, and Akchagylian stages (i.e. about 5.8–2 mya). Thus, these species may have subsequently immigrated into the Caspian depression from the Black Sea depression and other adjacent areas or are derived from taxa that did. We assume, after Mordukhai-Boltovskoi (1960), that post-Pliocene diversification in the Caspian fauna has been limited and largely occurred at the subspecies level. Recent evidence from three mt DNA sequences supports this hypothesis (Cristescu and Hebert 2002). However, a few Ponto-Caspian onychopod lineages including *Cornigerius* and *Podonevadne* have had a more prolonged episode of speciation from the Miocene to Pleistocene (Cristescu and Hebert 2002). Using palaeontological, ecological, and molecular evidence, we assembled a total of 456 extant

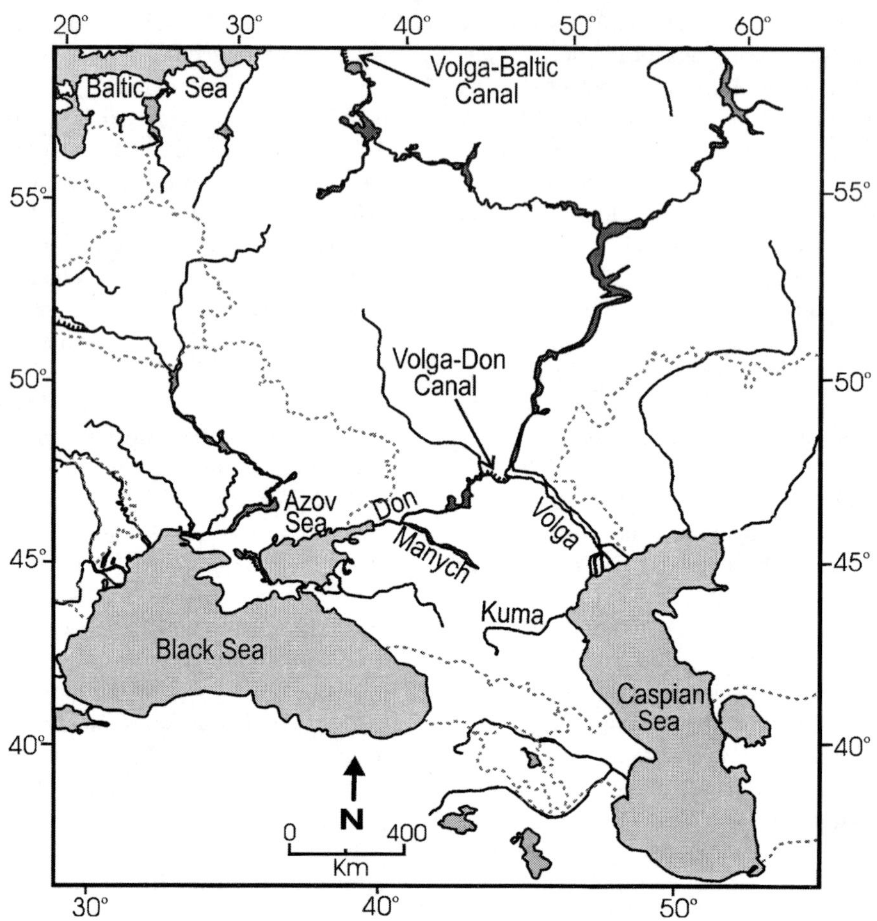

Figure 2. Map showing location of the modern Caspian, Black and Azov Seas and the Volga–Don Canal.

Table 2. Estimates of invertebrate species invasions to the Caspian basin during the last two million years (Late Pliocene–Holocene up to the early 1900s) via natural mechanisms.

Geologic epoch	Time (mya)	Source region	Number of species
Late Pliocene–Pleistocene	2.0–0.8	Black Sea basin	422
Pleistocene	1.8–0.01	Arctic basin	24
Holocene (up to the 1900s)	0.01–present	Azov–Black Sea basin	10

Summarized from Derzhavin 1951; Mordukhai-Boltovskoi 1960, 1979; Chepalyga 1985; M.E.A. Cristescu, unpub. data.

invertebrate species (Table 2) that naturally colonized the Caspian basin between the Late Pliocene and Holocene or are descended from Pliocene–Holocene non-Caspian ancestors; this approach is similar to that employed by Vermeij (1991). The bulk (92.5%) of these palaeobiological invasions is represented by Ponto-Caspian lineages that were classified by Mordukhai-Boltovskoi (1960, 1979). *Dreissena rostriformis* and *D. polymorpha* were not considered because they were present in older Akchagylian sediments (Zenkevich 1963). Arctic and Atlantic–Mediterranean lineages collectively represent 7.5% of invasions between the Late Pleistocene to Holocene, until the early 1900s (Table 2).

Mordukhai-Boltovskoi (1960) synthesized the long-term history of Ponto-Caspian lineages, in which periods of faunal isolations and species extinctions were intermingled with species invasions (see summary in Table 1). The precise timing of these faunal turnovers is poorly chronicled in the fossil record, except for some molluscan taxa (Mordukhai-Boltovskoi 1960). However, application of molecular clocks now provides a powerful tool for dating important historical processes shaping populations in isolated habitats such as past fragmentation and gene flow events.

The estimated number of Ponto-Caspian species involved in the Late Pliocene–Pleistocene colonization of the Caspian basin should be considered provisional because present species-level systematics remains controversial and species richness appears to be overestimated for several taxa (e.g. Mordukhai-Boltovskoi 1979; Grigorovich et al. 2000). For example, two morphologically distinctive members of the subgenera *Cercopagis* and *Apagis* (i.e. *Cercopagis* (*Cercopagis*) *pengoi* and *C.* (*Apagis*) *ossiani*) were characterized by the same haplotype for NADH dehydrogenase subunit 5 (ND5) mt gene in Lake Ontario, while allopatric European populations of *C. pengoi* consisted of up to five haplotypes (Makarewicz et al. 2001). These results suggest that *C. ossiani* and *C. pengoi* represent a single species, while the status of three other species of the subgenus *Apagis* await taxonomic re-assessment. Additional grounds for the inclusion of Ponto-Caspian taxa not chronicled in the fossil record were: (1) both non-molluscan and molluscan invertebrates often possess similar physiological tolerances, environmental preferences, and 'intrinsic invasive features' that could contribute equally to extinction or invasion; (2) many species are represented by divergent mtDNA conspecific lineages forming Caspian Sea and Black–Azov Sea clades that are morphologically almost indistinguishable (e.g. Cristescu et al. 2001; Cristescu and Hebert 2002).

Cercopagis lineages inhabiting the Caspian and Azov–Black Sea basins shared a common ancestry some 0.8 mya (Cristescu et al. 2001). Each surveyed population has experienced bottleneck events in the past as evidenced by limited genetic diversity at cytochrome *c* oxidase subunit 1 (COI) and ND5 loci (Cristescu et al. 2001). Arguably, Black Sea lineages of *Cercopagis pengoi* are characterized by higher genetic diversity than their Caspian counterparts, supporting Mordukhai-Boltovskoi's (1960) hypothesis that the former basin is the most likely source of the lineages in the Caspian basin. However, the basins currently support differing mtDNA lineages, suggesting the possibility that genetic drift has arisen from the founder effect.

By coupling mtDNA sequence diversity at COI, the ribosomal small and large subunits (12S and 16S) with the arthropod mt clock, Cristescu and Hebert (2002) estimated that generic and subgeneric differentiation within Onychopoda occurred over 10–20 mya. Although this estimate may be speculative due to problems with molecular clock calibration, it is congruent with the rate of genetic change estimated for Ponto-Caspian amphipod crustaceans (M.E.A. Cristescu, University of Guelph, Ontario, pers. comm.). Thus, the overall correspondence among taxa provides evidence for the antiquity of Ponto-Caspian lineages at genus and species levels (Richter et al. 2001; Cristescu and Hebert 2002).

There is some evidence that reduced gene flow with habitat isolation may produce intraspecific mtDNA divergence (e.g. Cristescu et al. 2001). However, available mtDNA sequence data on intraspecific populations residing in the Caspian and Azov–Black basins are limited to several crustacean species, including *Cercopagis pengoi*, *Cornigerius maeoticus*, *Podonevadne trigona*, *Pontogammarus maeoticus,* and *P. crassus* (M.E.A. Cristescu, pers. comm.). Surveys of mtDNA in these populations revealed limited genetic divergence at COI, rarely exceeding 2%, indicating that episodes of gene flow occurred between these lineages during the Pleistocene (Cristescu et al. 2001; M.E.A. Cristescu, pers. comm.). Thus, studies of mtDNA divergence provide new insights into the incidence of natural dispersal events that are poorly chronicled in the fossil record.

20th-century introductions of nonindigenous invertebrate species

Invasions occurred naturally in the Caspian Sea as a result of geological and climatic events throughout geologic time, while during historic time humans became the major vector for aquatic introductions (i.e. human-mediated invasions after Carlton 1996) in the Caspian region and throughout the world (Karpevich 1975; Ruiz et al. 1997; Cohen and Carlton 1998; Orlova 2000).

Our survey encompasses introduced nonindigenous species (NIS) that were absent from the Caspian Sea and Volga River delta prior to the 1900s, but which successfully colonized and naturalized in this region beyond their historic geographical ranges. Based on literature and field observations, we identified a total of 36 invertebrate NIS that were introduced through human-mediated vectors and became self-sustaining during the 20th century (Table 3).

Our ability to evaluate the extent of 20th-century invasions in the Caspian Sea is limited by the paucity of published information. Therefore, the number of recognized NIS appears underestimated and conservative compared to that recorded in other marine and estuarine

Table 3. Nonindigenous aquatic invertebrate species established in the Caspian Sea during the 20th century.

Taxonomic group	Species name	Year(s)	Native region	Entry vector	References
Mollusca	*Mytilaster lineatus*	1917–1919	Mediterranean	A	Bogachev 1928
Nematoda	Nematoda (NS)[1]	1917–1919?	Mediterranean	A	Chepalyga and Tarasov 1997
Crustacea	*Palaemon elegans*	1930–1934	Atlantic–Mediterranean	A	Zenkevich 1963
	Palaemon adspersus	1931–1934	Atlantic–Mediterranean	A	Zenkevich 1963
Crustacea	*Ergasilus* sp. (NS)[2]	1930–1934	Mediterranean–Pontic?	A	Mikailov 1958
Trematoda	*Haploporus longicolum*[2]	1930–1934	Mediterranean–Pontic	A	Mikailov 1958
	Haplosplanchnus pachysoma[2]	1930–1934	Mediterranean–Pontic	A	Mikailov 1958
	Saccocoelium obesum[2]	1930–1934	Mediterranean–Pontic	A	Mikailov 1958
	Ancyrocephalus vanbenedenii[2]	1930–1934	Mediterranean–Pontic	A	Zablotskii 1966
	Wlassenkotrema longicollum[2]	1930–1934	Mediterranean–Pontic	A	Zablotskii 1966
Polychaeta	*Nereis diversicolor*	1939–1940	Atlantic–Mediterranean	D	Hartman 1960
Mollusca	*Abra ovata*	1939–1940s	Atlantic–Mediterranean	D	Karpevich 1968
Crustacea	*Corophium volutator* f. *orientalis* (as *Corophium anodon*)	<1951	Atlantic–Mediterranean	?	Derzhavin 1951; Romanova 1975
	Balanus improvisus	1955	Pacific–Atlantic	S	Derzhavin 1956
	Balanus eburneus	1956	Atlantic	S	Zevina 1965
Coelenterata	*Blackfordia virginica*	1956	Pontic–West Atlantic	S	Logvinenko 1959
Crustacea	*Pleopis polyphemoides*	1957	Atlantic–Mediterranean	S	Mordukhai-Boltovskoi 1962
Bryozoa	*Electra crustulenta*	1958	Mediterranean	S	Abrikosov 1959
	Conopeum seurati	1958	Mediterranean	S	Zevina and Kuznetsova 1965
Crustacea	*Rhithropanopeus harrisi*	1958	West Atlantic	S	Nebolsina 1959
Mollusca	*Hypanis colorata*	1959	Pontic	M	Saenkova 1960
Coelenterata	*Moerisia maeotica*	1959	Pontic?	S	Mordukhai-Boltovskoi 1979
	Bougainvillia megas	1960	Pontic	S	Zevina 1965
Polychaeta	*Ficopomatus enigmaticus*	1961	Atlantic	S	Zevina 1965
Kamptozoa	*Barentsia benedeni*	1962	East Atlantic–Pacific	S	Zevina and Kuznetsova 1965
Crustacea	*Iphigenella shablensis*	<1969	Pontic	M?	Mordukhai-Boltovskoi et al. 1969
Mollusca	*Lithoglyphus naticoides*	1971	European	M	Pirogov 1972
Trematoda	*Apophallus muehlingi*	1971	European	M	Biserova 1990
	Rossicotrema donicum	1971	European	M	Biserova 1996
	Nicola scriabini	1971	European	M	Biserova 1996
Crustacea	*Acartia tonsa* (as *Acartia clausi*)	1981	Atlantic–Indian	S	Kurashova and Abdullaeva 1984
	Podon intermedius	1985	Atlantic–Mediterranean	S	Kurashova et al. 1992
Mollusca	*Tenellia adspersa*	1989	Atlantic–Mediterranean	S	Antsulevich and Starobogatov 1990
Crustacea	*Gammarus aequicauda*	<1994	Mediterranean	S	Piatakova and Tarasov 1996
Mollusca	*Dreissena bugensis*	1994	Pontic	S	Arakelova et al. 2000
Ctenophora	*Mnemiopsis leidyi*	<1999	West Atlantic	S	Ivanov et al. 2000

Year indicates time of first sighting or published record for inadvertently introduced species, and time interval for stocked target or contaminant species. Hosts of parasites: 1 – *Mytilaster lineatus*; 2 – *Mugil saliens*; 3 – *Lithoglyphus naticoides*. Keys to vectors of introduction: A – accidental releases; D – deliberate stocking; S – shipping activities; M – multiple vectors; ? – unknown or uncertain. NS – species identity not provided.

habitats of the world (see Ruiz et al. 1997, 2000). To test this possibility (i.e. that NIS introductions occurred but were not recognized), we sampled a series of shallow-water habitats in the Volga Delta and Caspian Sea in search of nonindigenous amphipod species. We selected the Amphipoda because they are the best-studied invertebrate group in the Caspian Sea, with 74 resident species recognized prior to commencement of maritime shipping via the Volga–Don seaway (Mordukhai-Boltovskoi 1960, 1979; Birshtein and Romanova 1968). During August 2000, samples were collected in the Volga Delta channels and littoral zone of the northern and central Caspian Sea, where several crustacean and molluscan NIS are known to occur. A dip net (250-mm mesh) was used for sweeping through vegetation, rocks, and bottom sediments. Using current taxonomic keys (Mordukhai-Boltovskoi et al. 1969; Grigorovich 1989), we identified two introduced amphipod species, *Iphigenella shablensis* and *Gammarus aequicauda* (= *Gammarus locusta*) that were previously reported, but were not recognized as established in the Caspian basin (see Piatakova and Tarasov 1996 for more details). However, our field survey revealed that both species are locally abundant (at densities >100 individuals m^{-2}), with *I. shablensis* occurring principally in the Volga Delta and northern Caspian and *G. aequicauda* in the central Caspian adjacent to the Apsheron peninsula. Populations of both *I. shablensis* and *G. aequicauda* included ovigerous females and juveniles, indicating that both species were reproducing and likely established.

Owing to uncertainties in systematics of groups with poorly distinguishable morphological features, the occurrence and sequence of NIS arrival are poorly documented for nematodes, polychaetes, and copepods (e.g. Chepalyga and Tarasov 1997). For example, the polychaete *Nereis diversicolor* was deliberately stocked in the Caspian Sea during 1939–1941, but was misidentified and treated as *Nereis succinea* for two decades (Hartman 1960). It is still not clear whether *N. succinea* is established in the Caspian Sea (e.g. Karpevich 1968, 1975). Likewise, the nonindigenous calanoid copepod *Acartia tonsa* has been erroneously reported as *A. clausi* (Kurashova and Abdullaeva 1984; N.V. Shadrin, Institute for Biology of Southern Seas, Sevastopol, Ukraine, pers. comm.). Detailed morphological analyses of calanoid copepods have revealed that the former species represents an Atlantic–Mediterranean introduction (in both the Caspian and Black Seas), while *A. clausi* is a Black Sea resident (N.V. Shadrin, pers. comm.).

During arrival and establishment phases, NIS are difficult to detect due to small population sizes. Several, or arguably most, introductions in the Caspian Sea were discovered only after populations had reached high abundances (e.g. the bivalve *Mytilaster lineatus*). For example, massive efforts were made to introduce the Atlantic–Mediterranean bivalve *Abra ovata* (=*A. segmentum*) between 1947 and 1948 (Chepalyga and Tarasov 1997). However, recent analyses of Caspian sedimentary deposits revealed that the species was inadvertently introduced and became established earlier, with stocked *Nereis*, between 1939 and 1941 (Chepalyga and Tarasov 1997). However, owing to a small population size, *A. ovata* remained undetected until 1955 (Karpevich 1968).

We are aware of three cyprinid fish species native to eastern Asia, grass carp *Ctenopharyngodon idella*, silver carp *Hypophthalmichthys molitrix*, and bighead carp *Aristichthys nobilis* that were stocked and became self-sustaining in many regions of Eurasia, including the Volga Delta and north Caspian Sea (Karpevich 1975; Ivanov 2000). One additional nontarget fish, snake-head *Ophiocephalus argus* was also introduced as a result of these stocking efforts (Zaitsev and Öztürk 2001). Globally these fishes are vectors for numerous parasitic helminthes (Karpevich 1975; Grigorovich et al. 2002). Because no record of parasites exists from the Caspian basin, it is possible that the estimate of the NIS fraction shall increase as additional data become available.

Nonindigenous species that have been successfully introduced to the Volga River delta and Caspian Sea during the 20th century include 58.3% of NIS native to the Atlantic–Mediterranean region (i.e. Lusitanian region of Atlantic Ocean, Mediterranean and Black Seas), 19.4% derived from the North Atlantic region (i.e. Atlantic coasts of Europe or North America), and 11.1% introduced from the Pontic region (i.e. inlets, coastal, and estuarine regions of northern Black Sea). Boreal European–Siberian region, Atlantic, Pacific, and Indian Oceans collectively account for the remaining 11.2% of NIS.

Nonindigenous invertebrates were introduced to the Caspian Sea through a variety of mechanisms, the importance of which varied over time (Figure 3A). Most of these introductions occurred after 1952, when the Caspian Sea was opened to maritime navigation through the Volga–Don Canal. Since that time, shipping has become the largest transport vector, accounting for 44.4% of all introductions to the Caspian Sea. Most sessile marine invaders arrived on ships' hulls during the late 1950s and early 1960s, whereas transportation in ballast tanks of cargo ships was likely the largest single entry mode for free-living invertebrates since the 1980s (Figure 3A, Table 3). During the 1970s, the combination of multiple and often interacting entry vectors (i.e. canal and commercial shipping traffic) facilitated entry of the gastropod *Lithoglyphus naticoides* and at least three parasitic trematode species carried with their gastropod host (Table 3). Thus, over a 40-year period since 1952, vectors related to shipping shifted from those associated with hull fouling to those associated with ballast water discharge (Grigorovich et al. 2002). Other NIS gained entry as nontarget species with aquaculture (27.8%) or through combinations of two or more entry mechanisms that acted jointly (19.4%). Two species (*Nereis diversicolor* and *Abra ovata*) were stocked deliberately. Introductions of invertebrates for commercial purpose started in 1897, although initial efforts to stock the Black Sea oyster *Ostrea edulis* and blue mussel *Mytilus galloprovincialis* were unsuccessful (Karpevich 1975).

Introductions of *Mytilaster lineatus*, *Abra ovata*, and *Lithoglyphus naticoides* are interesting because their fossil shells have been found in Caspian deposits

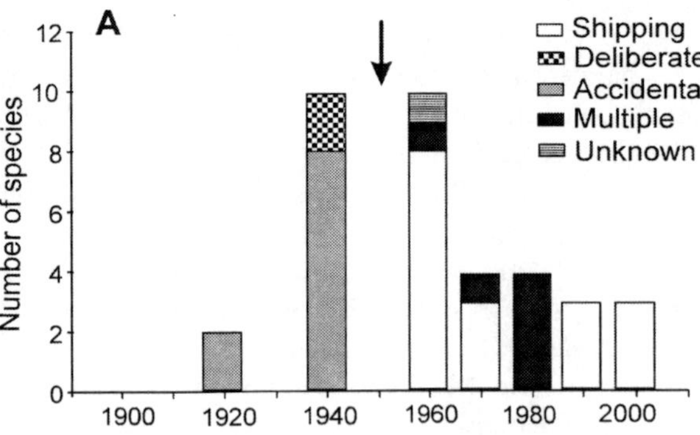

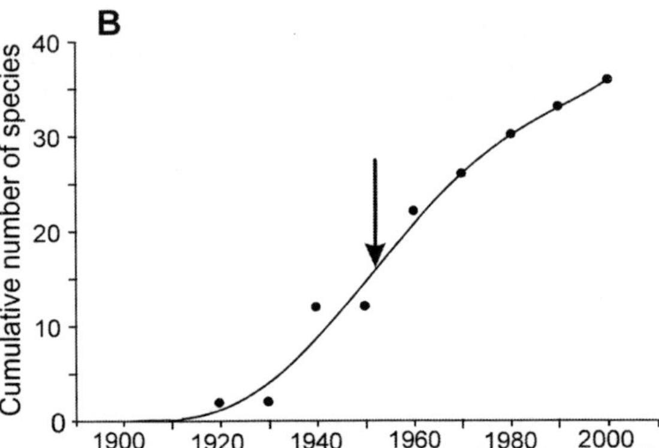

Figure 3. Time trends in introductions of nonindigenous aquatic invertebrates established in the Ponto-Caspian region, sorted by entry mechanism (A). Opening of the Volga–Don Canal (1952) is indicated by arrows. Cumulative number of nonindigenous invertebrate species established in the Caspian Sea and Volga River delta over the 20th century (B). Trends in nonindigenous invertebrate introductions are best described by polynomial function ($R^2 = 0.986$).

from earlier geologic epochs (Pirogov 1972; Orlova 2000). Consequently, the successful naturalization of these species appears to have resulted from multiple inoculations that occurred over an extended period of time.

Invasion trends in the Caspian Sea

Species invasions over the last two million years have challenged the biological integrity of the Caspian Sea. Long-term fluctuations in sea level and climate have breached natural barriers to dispersal for species from several donor regions, notably Lake Pannon, the Black Sea basin, and the Arctic Ocean. During the 20th century, several invertebrate invasions in the Volga Delta and adjacent northern Caspian Sea likely resulted from natural passive dispersal through wind or waterfowl. Examples include the anomopod cladoceran *Daphnia lumholtzi*, which is indigenous to Australia, southern Asia, and Africa, and the bryozoan *Lophopodella carteri*, which is native to southern Asia, Indonesia, and Africa (Benhing 1928; Abrikosov and Kosova 1963). However, during the 20th century (Figure 3B), human-mediated invasions have occurred with increasing frequency and now constitute the dominant mechanism of the Caspian Sea fauna change (Logvinenko 1965; Ivanov et al. 2000). Recent human-mediated invasions are attributable largely to the creation of 'invasion corridors' (i.e. primary transportation vectors and dispersal pathways) linking the Caspian Sea with the Azov and Black Seas (Ricciardi and MacIsaac 2000; MacIsaac et al. 2001). Since the opening of the Volga–Don Canal in 1952, shipping traffic has become the largest vector for transfer of NIS to the Caspian Sea (Figure 3A). Among the 36 NIS recorded in the Caspian Sea and Volga Delta after 1952, 23 taxa possess a suite of characteristics amenable to transportation on ship hulls or in ballast tanks (Zevina 1965; Zevina and Kuznetsova 1965; Gollasch and Leppäkoski 1999).

A recent series of invasions by *Mnemiopsis leidyi* and other marine invertebrates indicates that ballast water transport has become the largest vector for NIS transfer to the Caspian Sea. Three factors that mediate the rate of marine invasions are: (1) an increase in vector size (e.g. more ballast water due to more or larger ships); (2) an increase in NIS inoculation owing to faster ships; and (3) an improvement in vector quality (e.g. ballast water is not contaminated with pollutants) (Carlton 1996). Future NIS introductions are expected to continue as long as expanding regional and global trade provides enhanced dispersal opportunities for aquatic organisms (Gollasch and Leppäkoski 1999; Ivanov 2000).

Based on combined molecular, palaeontological, and ecological data, we recognize 456 invertebrate species that have, since the Late Pliocene, naturally colonized the Caspian basin or are descended from the these taxa. Thus, over the past two million years, the long-term average rate of species invasions is approximately 0.02 species per 100 years. In contrast, the rate of human-mediated establishment is 36 species over the past 100 years. Therefore, the incidence of human-mediated species introductions in the Caspian Sea corresponds to an 1800-fold increase in rate of invasion compared to the background, natural level. Although natural dispersal vectors may account for some recent invasions, human-mediated transfer is dominant, due primarily to shipping activities (Carlton 1996). Hebert and Cristescu (2002) estimated that human-mediated dispersal of cladocerans might exceed the natural rate in the Laurentian Great Lakes by up to 50,000-fold.

During the last two million years, the Caspian basin has been a target and donor region during several episodes of faunal interchange via natural connections that were temporarily established with the Black Sea basin and the Arctic Ocean. Species and possibly higher-level clades have dispersed to the Caspian Sea and naturalized. Interestingly, this faunal interchange has become strongly asymmetrical since the Holocene. Numerous NIS have been transferred to the Caspian Sea from the Black–Azov basins and from other regions around the world (Tables 2 and 3), yet no invertebrates have dispersed in the opposite direction (e.g. Nikolaev 1979).

Currently, over 30 Ponto-Caspian metazoan species reside in the Volga, Ural and other southward flowing rivers that empty into the Caspian Sea (Mordukhai-Boltovskoi 1960, 1979). Geographical patterns of these distributions are consistent with pathways provided by the rivers, which flow in north-south directions. Arguably, natural barriers (i.e. strong currents, salinity, and ion content) historically served to limit the dispersal by Ponto-Caspian invertebrates in these potential 'invasion corridors' (Mordukhai-Boltovskoi 1960). Many of these distributional patterns have been interpreted as the result of dispersal events, principally those mediated by humans during the last century (Dumont 1998, 2000; Orlova 2000). However, recent morphological and molecular evidence provides little support for this hypothesis. For example, at least two invading onychopod cladocerans endemic to the Black–Azov basin, *Podonevadne trigona ovum* and *Cornigerius maeoticus maeoticus*, were identified for the first time from the Volgograd reservoirs during the 1990s (I.K. Rivier, Institute for Inner Water Biology, Borok, Russia, pers. comm.), while the occurrence of phenotypically-divergent, conspecific forms endemic to the Caspian Sea (i.e. *P. trigona trigona* and *C. maeoticus hircus*) have not been confirmed in this reservoir (I.K. Rivier, pers. comm.). Likewise, molecular markers have revealed that the Black–Azov basin, not the Caspian Sea, was the source of invasion of the Volgograd Reservoir by *Cercopagis pengoi* (M.E.A. Cristescu, pers. comm.).

Application of fossil and molecular evidence into the study of historical invasions provides insights on the accelerating rate of species invasion in the Caspian Sea that has been mediated by anthropogenic activities. However, additional fossil and molecular data will be required to provide finer resolution of the invasion history of this basin and the role of human activities in shaping the current Caspian fauna.

Acknowledgements

I.K. Rivier provided information on Ponto-Caspian onychopod cladocerans in Volgograd Reservoir, Russia. M.E.A. Cristescu provided mtDNA information. Comments by J. Pederson, N. Macdonald, D. Berg, V. Anistratenko, M. Weis, M. Docker, C. Busch, H. Limén, S.A. Bandoni and two anonymous reviewers improved the manuscript. Financial support from the Natural Sciences and Engineering Research Council of Canada (NSERC) and Premier's Research Excellence Award Program (to H.J.M.), a GLIER postdoctoral fellowship (to I.A.G.), and an NSERC postdoctoral fellowship (to T.W.T) are gratefully acknowledged.

References

Abrikosov GG (1959) A new invader in the Caspian Sea. Zoologicheskii Zhurnal 38: 1745–1746 [in Russian]

Abrikosov GG and Kosova AA (1963) On finding of the tropical freshwater bryozoan *Lophopodella carteri* (Bryozoa, Phylactolaemata). Zoologicheskii Zhurnal 42: 1724–1726 [in Russian]

Antsulevich AE and Starobogatov YI (1990) First finding of the nudibranch mollusc (Order Tritoniiformes) in the Caspian Sea. Zoologicheskii Zhurnal 69: 138–140 [in Russian]

Arakelova ES, Orlova MI and Filippov AA (2000) Hydrobiological research carried out by the Zoological Institute of the RAS in the Volga delta and in the northern part of the Caspian Sea during 1994–1997. Caspian Floating University. Research Bulletin 1: 102–108 [in Russian]

Benhing AL (1928) Studien ueber die Crustaceen des Wolgabassins. Archiv für Hydrobiologie 19: 423–432

Berg DJ and Garton DW (1994) Genetic differentiation in North American and European populations of the cladoceran *Bythotrephes*. Limnology and Oceanography 39: 1503–1516

Berg DJ, Garton DW, MacIsaac HJ, Panov VE and Telesh IV (2002) Changes in genetic structure of North American *Bythotrephes* populations following invasion from Lake Ladoga, Russia. Freshwater Biology 47: 275–282

Birshtein YA and Romanova NN (1968) Order Amphipoda. In: Birshtein YA, Vinogradov LG, Kondakov NN, Astakhova MS and Romanova NN (eds) Atlas of Invertebrates of the Caspian Sea, pp 241–289. Pishchevaya Promyshlennost, Moscow [in Russian]

Biserova LI (1990) Occurrence and distribution of *Lithoglyphus naticoides* (Gastropoda, Lithoglyphiidae) in the Volga delta. Gidrobiologicheskii Zhurnal 26(2): 98–100 [in Russian]

Biserova LI (1996) Parasites of the invading mollusc *Lithoglyphus naticoides* (C. Pfr.) in the Volga River delta. In: Problems of Hydrobiology of Continental Waters and Their Molluscan Fauna. Abstracts in honour of the centennial birthday of Prof V.I. Zhadin, pp 13–14. Zoologicheskii Institut Rossiiskoi Akademii Nauk, St Petersburg [in Russian]

Bogachev VV (1928) *Mytilaster* in the Caspian Sea. Russkii Gidrobiologicheskii Zhurnal 7(8–9): 187–188 [in Russian]

Carlton JT (1996) Patterns, process, and prediction in marine invasion ecology. Biological Conservation 78: 97–106

Chepalyga AL (1985) Inland sea basins. In: Velichnko AA (ed) Late Quaternary Environments of the Soviet Union, pp 229–247. University of Minnesota Press, Minneapolis, Minnesota

Chepalyga AL and Tarasov AG (1997) Introduction of Atlantic species into the Caspian Sea: the fate of endemic taxa and biosystems. Okeanologiya 37: 261–268 [in Russian]

Cohen A and Carlton JT (1998) Accelerating invasion rate in a highly invaded estuary. Science 279: 555–558

Cristescu MEA and Hebert PDN (2002) Phylogeny and adaptive radiation in the Onychopoda (Crustacea, Cladocera): evidence from multiple gene sequences. Journal of Evolutionary Biology 15: 838–849

Cristescu MEA, Hebert PDN, Witt JDS, MacIsaac HJ and Grigorovich IA (2001) An invasion history for *Cercopagis pengoi* based on mitochondrial gene sequences. Limnology and Oceanography 46: 224–229

Derzhavin AN (1951) Study in the history of the fauna of the Caspian and freshwater basins of Azerbaijan. In: Alizade AN, Asadov SM and Derzhavin AN (eds) Animal World of Azerbaijan, pp 34–67. Izdatelstvo Akademii Nauk Azerbaidzhanskoi SSR, Baku [in Russian]

Derzhavin AN (1956) New invader in the Caspian Sea – the marine barnacle *Balanus improvisus* Darwin. Doklady Akademii Nauk Azerbaijanskoi SSR 12(1): 43–47 [in Russian]

Dumont HJ (1998) The Caspian lake: history, biota, structure, and function. Limnology and Oceanography 43: 44–52

Dumont HJ (2000) Endemism in the Ponto-Caspian fauna, with special emphasis on the Onychopoda (Crustacea). Advances in Ecological Research 31: 181–196

Geary DH, Magyar I and Müller P (2000) Ancient Lake Pannon and its endemic molluscan fauna (Central Europe; Mio-Pliocene). Advances in Ecological Research 31: 463–482

Gollasch S and Leppäkoski E (1999) Initial risk assessment of alien species in Nordic coastal waters. Nord 1999: 8. Nordic Council of Ministers, Copenhagen, Denmark

Grigorovich IA (1989) Guide for identification of amphipods of fresh and brackish waters of the southwestern USSR. Institute of Hydrobiology, Academy of Sciences of the Ukrainian SSR, Kiev [in Russian]

Grigorovich IA, MacIsaac HJ, Rivier IK, Aladin NV and Panov VE (2000) Comparative biology of the predatory cladoceran *Cercopagis pengoi* from Lake Ontario, Baltic Sea and Caspian Sea. Archiv für Hydrobiologie 149: 23–50

Grigorovich IA, MacIsaac HJ, Shadrin NV and Mills EL (2002) Patterns and mechanisms of aquatic invertebrate introductions in the Ponto-Caspian region. Canadian Journal of Fisheries and Aquatic Sciences 59: 1189–1208

Hartman O (1960) An account of the nereid worms, *Neanthes diversicolor*, a new combination in the Caspian Sea and its more extensive distribution. Zoologicheskii Zhurnal 39: 35–39 [in Russian]

Hebert PDN and Cristescu MEA (2002) Genetic perspectives on invasions: the case of the Cladocera. Canadian Journal of Fisheries and Aquatic Sciences 59: 1229–1234

Ivanov VP (2000) Biological resources of the Caspian Sea. KaspNIRKh imprint, Astrakhan [in Russian]

Ivanov VP, Kamakin AM, Ushivtzev VB, Shiganova T, Zhukova O, Aladin N, Wilson SI, Harbison GR and Dumont HJ (2000) Invasion of the Caspian Sea by the comb jellyfish *Mnemiopsis leidyi* (Ctenophora). Biological Invasions 2: 255–258

Karpevich AF (ed) (1968) Acclimatization of Fish and Invertebrates in the Waterbodies of the USSR. Nauka, Moscow [in Russian]

Karpevich AF (1975) Theory and practice of acclimatization of aquatic organisms. Pishchevaya Promyshlennost, Moscow [in Russian]

Kasymov AG (1982) The role of Azov–Black Sea invaders in the productivity of the Caspian Sea benthos. Internationale Revue der gesamten Hydrobiologie 67: 533–541

Kasymov AG (1987) The Animal Kingdom of the Caspian Sea. Elm, Baku [in Russian]

Kurashova EK and Abdullaeva NM (1984) *Acartia clausi* (Calanoida, Acartiidae) in the Caspian Sea. Zoologicheskii Zhurnal 63: 931–933 [in Russian]

Kurashova EK, Tinenkova DKh and Yelizarenko MM (1992) *Podon intermedius* (Cladocera, Podonidae) in the Caspian Sea. Zoologicheskii Zhurnal 71: 135–137 [in Russian]

Logvinenko BM (1959) On finding the medusa *Blackfordia virginica* in the Caspian Sea. Zoologicheskii Zhurnal 38: 1257–1258 [in Russian]

Logvinenko BM (1965) On the alterations in the fauna of the Caspian molluscs of the genus *Dreissena* after invasion by *Mytilaster lineatus* (Gmel.). Nauchnye doklady vysshei shkoly. Biologicheskiye Nauki 4: 14–19 [in Russian]

MacIsaac HJ, Grigorovich IA and Ricciardi A (2001) Reassessment of species invasion concepts: the Great Lakes basin as a model. Biological Invasions 3: 405–416

Makarewicz JC, Grigorovich IA, Mills EL, Damaske E, Cristescu ME, Pearsall W, LaVoie MJ, Keats R, Rudstam L, Hebert PDN, Halbritter H, Kelly T, Matkovich C and MacIsaac HJ (2001) Distribution, fecundity, and genetics of *Cercopagis pengoi* (Ostroumov) (Crustacea: Cladocera) in Lake Ontario. Journal of Great Lakes Research 27: 19–32

Mikailov TK (1958) Parasitic fauna of the Caspian Sea mullets. Zoologicheskii Zhurnal 37: 373–378 [in Russian]

Mordukhai-Boltovskoi FD (1960) Caspian fauna in the Azov and Black Sea Basin. Izdatelstvo Akademii Nauk SSSR, Moscow [in Russian]

Mordukhai-Boltovskoi FD (1962) Appearance of the Mediterranean polyphemid in the Caspian Sea. Zoologicheskii Zhurnal 41: 289–290 [in Russian]

Mordukhai-Boltovskoi FD (1964) Caspian fauna beyond the Caspian Sea. Internationale Revue der gesamten Hydrobiologie 49: 139–176

Mordukhai-Boltovskoi FD (1979) Composition and distribution of Caspian fauna in the light of modern data. Internationale Revue der gesamten Hydrobiologie 64: 1–38

Mordukhai-Boltovskoi FD, Greze II and Vasilenko SV (1969) Order Amphipoda. In: Vodianitskii VA (ed) Guide for Identification of the Fauna of the Black and Azov Seas. Freeliving Invertebrates, Vol 2, Crustaceans, pp 440–524. Naukova Dumka, Kiev [in Russian]

Nebolsina TK (1959) Crab in the Caspian Sea. Priroda (Moscow) 6: 116–117 [in Russian]

Nikolaev II (1979) Ecological consequences of unintended anthropogenic distribution of aquatic fauna and flora. In: Smirnov NN (ed) Ecological Prognostication, pp 76–93. Nauka, Moscow [in Russian]

Orlova MI (2000) The Caspian basin as a region-donor and a region-recipient of bioinvasions of water invertebrates. In: Matishov GG (ed) Species Introducers in the European Seas in Russia, pp 58–75. Kol'skii Nauchnyi Tsentr RAN, Apatity [in Russian]

Pirogov VV (1972) On the finding of *Lithoglyphus naticoides* in the Volga delta. Zoologicheskii Zhurnal 51: 912–913 [in Russian]

Piatakova GM and Tarasov AG (1996) Caspian Sea amphipods: biodiversity, systematic positions and ecological peculiarities of some species. International Journal of Salt Lake Research 5: 63–79

Rass TS (1978) Introductions and acclimatizations of marine organisms, their parameters and importance. In: Vinberg GG (ed) Elements of Aquatic Systems. Trudy Vsesoyuznogo Gidrobiologicheskogo Obshchestva 22, pp 70–94. Nauka, Moscow [in Russian]

Ricciardi A and MacIsaac HJ (2000) Recent mass invasion of the North American Great Lakes by Ponto-Caspian species. Trends in Ecology and Evolution 15: 62–65

Richter S, Braband A, Aladin N and Scholtz G (2001) The phylogenetic relationships of 'predatory water-fleas' (Cladocera: Onychopoda, Haplopoda) inferred from 12S rDNA. Molecular Phylogenetics and Evolution 19: 105–113

Romanova NN (1975) Quantitative distribution and ecology of corophiids (Crustacea, Amphipoda, *Corophium*). Byulleten Moskovsko obshchestva ispytatelei prirody. Otdel Biologicheskii 80(3): 51–63 [in Russian]

Ruiz GM, Carlton JT, Grosholz ED and Hines A (1997) Global invasions of marine and estuarine habitats by nonindigenous species: mechanisms, extent, and consequences. American Zoologist 37: 621–632

Ruiz GM, Fofonoff PW, Carlton JT, Wonham MJ and Hines AH (2000) Invasion of coastal marine communities in North America: apparent patterns, processes, and biases. Annual Review of Ecology and Systematics 31: 481–531

Saenkova AK (1960) *Monodacna colorata* in the Caspian Sea. Priroda (Moscow) 11: 111 [in Russian]

Sovinskii VK (1904). An introduction to the study of the fauna in the Ponto-Caspian-Aral marine basin considered as an independent zoogeographical province. Zapiski Kievskogo obshchestva Estestvoispytatelei 18: 1–487, plates 1–4 [in Russian]

Stepien CA, Hubers AN and Skidmore JL (1999) Diagnostic genetic markers and evolutionary relationships among invasive dreissenid and curbicolid bivalves in North America: phylogenetic signal from mitochondrial 16S rDNA. Molecular Phylogenetics and Evolution 13: 31–49

Vermeij GJ (1991) Anatomy of an invasion: the trans-Arctic interchange. Paleobiology 17: 282–307

Zablotskii VI (1966) Changes in the parasite fauna of mullet in connection with its acclimatization in the Caspian Sea. In: Yanushevich AI (ed) Acclimatization of Animals in the USSR, pp 238–239. Israel Program for Scientific Translations, Jerusalem

Zaitsev Y and Öztürk B (eds) (2001) Exotic Species in the Aegean, Marmara, Black, Azov and Caspian Seas. Turkish Marine Research Foundation, Istanbul

Zenkevich LA (1963) Biology of the Seas of the USSR. Interscience Publishers, New York

Zevina GB (1965) Fouling biocenoses in the Caspian Sea and their alterations associated with introduction of new organisms. In: Zenkevich LA (ed) Change of Biological Assemblages of the Caspian Sea during the Last Decade. Nauka, Moscow [in Russian]

Zevina GB and Kuznetsova IA (1965) The role of shipping in the alteration of the Caspian Sea fauna. Okeanologiya 5: 518–527 [in Russian]

Biological Invasions **5**: 117–131, 2003.

Biological control of marine invasive species: cautionary tales and land-based lessons

David Secord
University of Washington, Tacoma, IAS Environmental Science Program, 1900 Commerce St., Mailstop 358436, Tacoma, WA 98402-3100, USA (e-mail: dave@u.washington.edu; fax: +1-253-692-5718)

Received 30 August 2001; accepted in revised form 23 April 2002

Key words: algae and seagrasses, alien or exotic species, biocontrol of marine pests, biological invasions of marine invertebrates, host range expansion, host shifts, host specificity paradigm, natural enemies, non-indigenous species, predator–prey interactions, risk and conservation biology

Abstract

Biological control (biocontrol) has successfully regulated pest populations in terrestrial agroecosystems, but it has also caused negative unintended consequences for native species. Marine biologists and resource managers have recently published a growing number of proposals to include biocontrol in integrated pest management programs in oceans, seas and estuaries. Here, I review six ecologically and taxonomically diverse case studies of marine biocontrol programs at various stages of planning and implementation. Proposals include viral or microbial control of harmful algal blooms, predatory control of the ctenophore *Mnemiopsis leidyi* in the Black Sea, parasitic regulation of the European green crab *Carcinus maenas*, castration by ciliates of the seastar *Asterias amurensis* in Australia, herbivory of the toxic green alga *Caulerpa taxifolia* in the Mediterranean by sacoglossan sea slugs, and insect biocontrol by the planthopper *Prokelesia marginata* to ameliorate ecological impacts of the saltmarsh cordgrass *Spartina alterniflora*. Where data exist, I evaluate these examples in terms of lessons marine invasion biologists can glean from the rich history of terrestrial biocontrol, and explicitly contrast agroecosystems with invaded marine habitats. Host specificity cannot be guaranteed in the marine biocontrol proposals examined. Feasible alternatives to classical biocontrol in the marine realm should be emphasized, including more investment in invasion prevention tools, early detection and eradication while invasions are small, and increased attention to native natural enemies to control exotic pests. Biocontrol in marine habitats is risky: it poses many more uncertainties and has a much sparser history than its counterpart on land.

Abbreviations: HSP – host specificity paradigm; HS – host switching; HRE – host range expansion; Biocontrol – biological control

Introduction: successes and pitfalls of biological control on land

Biological control, or biocontrol, is the reliance upon natural enemies to attack pest organisms that damage human interests. Typically, perceived pest species are not native to the region where they cause problems; however, organisms used to regulate pest populations may be either native or introduced. In this paper 'natural enemies' connotes exploiter–victim species' interactions in an ecological sense, without necessarily implying strict coevolution of species pairs (Crawley 1992; Thompson 1994). Controlling pests by using other organisms to eat, parasitize, infect, or compete with them is a tradition that informally dates back many centuries. Large-scale biocontrol in terrestrial agricultural ecosystems, though, is generally recognized as having its genesis in 1888 with the successful control in California of cottony-cushion scale, a citrus pest, via the introduction of the vedalia beetle from Australia (Doutt 1958; Sweetman 1958). In this paper I review some terrestrial biocontrol history to establish a context for evaluating several recent proposals for biological pest control in the marine/estuarine realm, which differs from that on land in several critical respects germane to the safety and efficacy of biocontrol. Important contrasting attributes of marine ecosystems include organisms' life histories and taxonomic identities, typical larval or adult dispersal strategies, community interactions and biodiversity, biomechanics of the medium, and relative openness of the system (Strathmann 1990).

Biocontrol is generally considered to take several distinct forms. These include *classical biocontrol*, in which a natural enemy from the pest's native range is introduced to control it in its new range. *Neoclassical biocontrol* is the controversial practice of introducing a non-indigenous species to control a *native* pest, on the grounds that 'new associations' between antagonist and victim species are more likely to be effective in the context of pest control (Hokkanen and Pimentel 1984). Finally, *augmentative biocontrol* is the practice of enhancing populations of predators, parasites, or pathogens to improve their regulation of native or introduced pests. This may include enhancement of native enemies' populations or habitats through *indigenous biocontrol* or *conservation biocontrol* (Settle et al. 1996; Van Driesche and Bellows 1996; Barbosa 1998; Chang and Kareiva 1999). In the cases of classical and neoclassical biocontrol, another non-indigenous species is deliberately introduced to control a pest. In all these types of biocontrol programs, practitioners try to take advantage of a negative species interaction in an ecological interaction web to reduce survivorship or reproduction of an 'undesirable' species. The goal is usually not to eliminate the pest species altogether, but rather to keep it sustainably suppressed at socially or economically acceptable levels (Gause 1969). Total eradication, while desirable, is generally not considered feasible once an introduced species has become established (Van Driesche and Bellows 1996).

Since the dramatic 19th-century reduction of cottony-cushion scale in California, biocontrol has achieved some spectacular successes on land but has also produced some high-profile catastrophes. Many of these failures were unintended consequences of the introduction of non-native biocontrol agents, with their subsequent attacks on economically or ecologically valuable non-target organisms (Clarke et al. 1984; Howarth 1991; Simberloff and Stiling 1996a, b). Indeed, for its first several decades biological control was rather 'hit and miss', a process by which biologists explored the native range of a pest to find potential natural enemies that might be imported and deployed against it. In this way, hundreds of species were introduced to all parts of the world, many simultaneously rather than one at a time, and usually by trial and error without carefully controlled experiments (Sweetman 1958; Van Driesche and Bellows 1996).

Today, decisions on biocontrol agent identification and release are usually made more deliberately, with forethought typically guided by two key principles: host specificity and maximizing damage to the pest. Specific regulations for testing and release, however, vary by country, by taxon, by habitat, and by trophic role of the proposed biocontrol agent. In agroecosystems, this first selection principle is typically addressed by means of 'host specificity testing' designed to gauge the range of hosts that might be attacked or infected by a potential biocontrol species. Following many notoriously destructive biocontrol introductions of generalist predators such as the mongoose and rosy wolf snail in Hawaii and the cane toad in Australia, the 'host specificity paradigm' (HSP) is now the dominant framework governing how potential biocontrol species are selected (Blossey 1995; Secord and Kareiva 1996). The latter principle, host damage, is important because even specialized consumers may have minimal or highly variable demographic effects on their hosts and so would not make good biocontrol agents (see Bigger and Marvier 1998). For example, many sacoglossan sea slugs are monophagous or oligophagous without being ecologically important to their algal hosts (Williams and Walker 1999). It is now generally recognized that a successful biocontrol agent must not only damage its target, but must do so in a host specific way.

The host specificity paradigm in biological control

The key principle behind the host specificity paradigm is to find a biocontrol agent that will inflict maximal damage on the pest while doing minimal damage to anything else (non-target effects on non-target organisms). Biocontrol agents may be considerably more specific than chemical pesticides in what they harm, which is one of the reasons they are widely considered environmentally 'safe and clean' (e.g. Doutt and Smith 1971; Simberloff and Stiling 1996a, b). Unlike most chemical pesticides, though, their ecological effects are almost always irreversible; once they are established in a new environment, they are typically there to stay. Moreover, as dynamic populations of living organisms, they are free to disperse and to evolve, often rapidly in ecological time; such evolutionary changes are often difficult to predict *a priori* (Secord and Kareiva 1996; Ehler 1998). Even some famous biocontrol success stories are mixed, such as the introduction of *Cactoblastis* moths to control *Opuntia* cacti; while moths successfully controlled the cacti in Australia, they have dispersed on their own and attacked endangered cactus species after being introduced to Caribbean islands (Simberloff 1992; Cory and Myers 2000). Finally, non-target effects often involve indirect interactions which are not detected by conventional host specificity testing (see chapters in Follett and Duan 2000; Barratt et al. 2001; Holt and Hochberg 2001; Waage 2001). A good example is the competitive replacement of most native ladybird beetles in North America with the semi-specialized predatory European species introduced to control aphids in agroecosystems (Elliott et al. 1996).

Recently, some concerns have arisen regarding the environmental safety of the HSP (Secord and Kareiva 1996; Simberloff and Stiling 1996a, b; Ehler 1998; Cory and Myers 2000). Unlike chemical pesticides, biological control agents are reproductive, genetically variable populations of organisms. Thus, they display phenotypic plasticity as well as evolutionary responses to strong selection in novel or changing environments. Such conditions are likely to prevail in the introduced ranges of biocontrol agents. Indeed, in the intimate interspecific associations often considered desirable for biocontrol, hosts are a significant component of the pest's environment exerting especially strong selection pressure (Carroll and Boyd 1992; Brooks and McLennan 1993; Secord and Kareiva 1996; Secord 2001). When a species that customarily attacks only one host species shifts its exclusive affinities to another host species (or set of species), it is engaging in host switching (HS). When a species *adds* another target species to its repertoire of potential hosts, this is known as host range expansion (HRE). Either way, species not part of the original host range become non-target hosts of the biocontrol agent (recent insect examples include Carroll and Boyd 1992; Dennill et al. 1993; Hawkins and Marino 1997; Louda et al. 1997; Boettner et al. 2000; Henneman and Memmott 2001). Host changes like these are unlikely to have been detectable *a priori* by host specificity screening in the laboratory, especially if they involve substantial morphological or behavioral evolution by natural enemies in ecological time (see Carroll and Boyd 1992 for an exceptionally well-documented example).

Biocontrol is now widely considered a critical component of many integrated pest management (IPM) programs (Dent 1995). Such programs are beginning to spread from the land to the water due to increases in the numbers and impacts of aquatic bioinvasions. Several proposals for freshwater animal (zebra mussels; Molloy 1998) and plant (Eurasian watermilfoil, water hyacinth, purple loosestrife, and *Hydrilla*; Habeck 1983; Blossey et al. 1994a, b; Balciunas et al. 1996; Stanley and Julien 1999) biocontrol have recently been proposed or attempted. At least one multispecies aquatic biocontrol program, the introduction of leaf beetles and weevils to control purple loosestrife in North America, has been strongly criticized on the grounds that the impacts of the initial invasion have been exaggerated and thus do not warrant the introduction of non-indigenous insects for its control (Hager and McCoy 1998). Hager and McCoy (1998) suggest that irreversible biocontrol releases may have been undertaken based on inadequate experimental evidence and untested hypotheses. That said, we likely know more about the population dynamics and demographic details of the purple loosestrife invasion than we do about *any* of the marine invasions profiled here.

Even disregarding early introductions of generalist predators to island ecosystems that most now recognize as mistakes, concerns remain about classical biocontrol in this era of routine host specificity testing. Chief among these concerns are (1) indirect effects of a biocontrol agent on non-target organisms; (2) evolution of host affinities (HS or HRE) due to strong selection, genetic variation, and novel ecological opportunities to exploit new hosts; and (3) dispersal of biocontrol agents that puts them into contact with new, native, endangered or commercially valuable hosts. Non-target effects of classical biological control agents are increasingly being discovered (Howarth 1991 and sources cited therein; Dennill et al. 1993; Secord and Kareiva 1996 and sources cited therein; Simberloff and Stiling 1996a, b and sources cited therein; Hawkins and Marino 1997; Louda et al. 1997; Boettner et al. 2000; Henneman and Memmott 2001). Moreover, in an illustration of the adage 'absence of evidence is not evidence of absence', biocontrol practitioners often assume that ecological impacts on non-target organisms do not exist, simply because no attempts have been made to detect such effects (Simberloff 1992). For example, what are the consequences for unsurveyed natural areas 500 m, 5 or 50 km from the large, highly simplified agroecosystems where biocontrol species have been introduced? Given marine currents and larval dispersal patterns, such uncertainties would be greatly magnified in biocontrol at sea. In those cases where biologists have looked for non-target effects, they have often found them (Simberloff 1992; Simberloff and Stiling 1996a, b; Louda et al. 1997; Hawkins and Marino 1997; Henneman and Memmott 2001). Furthermore, biocontrol advocates have suggested that the laboratory or greenhouse conditions under which host specificity testing usually occurs indicate a broader range of host species for a prospective biocontrol agent than would be found in the field. However, when field and lab host ranges were explicitly compared for biocontrol species (two weevil-attacking parasitoids in New Zealand), the field host range reflected that detected in the lab, and for one of the biocontrol agents it included a wide diversity of native weevil species (Barratt et al. 1997).

Historically not part of the management of marine invasive species, biocontrol as part of marine IPM is now being increasingly proposed, discussed, and even implemented. Therefore, this is an appropriate time for a review of marine biocontrol, in light of what we have learned from over a century of formal large-scale implementation of terrestrial biocontrol.

Six case studies of marine biocontrol

To illustrate the potential and pitfalls of transferring biocontrol strategies and technologies to the marine realm, I describe in detail six recent proposals to use natural enemies in estuaries and oceans. Each presents unique biological and ethical quandaries that are in many ways distinct from – and less well understood than – those presented by the history of terrestrial biocontrol, with its agricultural context and applications. Indeed, it is difficult to overemphasize the extreme differences in the spatial aspects and species richness of agricultural ecosystems in contrast to most marine ones. I conclude by evaluating these marine cases in light of the general prospects for marine biocontrol – including alternative strategies – that might become durable solutions to the problem of prevention and control of marine invasive species.

Biocontrol of harmful algal blooms

Mounting global concern over increased frequency of toxic or harmful algal blooms (HABs) has led to the search for evidence that phytoplankton, like many other marine taxa, may be transported in ships' ballast water (Carlton 1999; McCarthy and Crowder 2000). Such algal blooms may not be merely a function of changing environmental conditions, but also of new invasive algal species arriving in different geographic regions (McCarthy and Crowder 2000). Recent evidence that viruses may naturally play significant ecological roles in regulating phytoplankton blooms, including HABs (e.g. Bratbak et al. 1993; Milligan and Cosper 1994) has produced proposals to use phytoplankton-attacking microorganisms (viruses, bacteria, or protozoans) as biological control agents (Nagasaki et al. 1999).

Viruses attack eukaryotic HAB-forming *Heterosigma* cells by causing them to lyse, and are often associated with the decay phase of blooms (Bratbak et al. 1993; Brussaard et al. 1996; Nagasaki et al. 1999). Algicidal bacteria have also been implicated in terminating *Heterosigma* blooms in Japan (Kim et al. 1998). In the understudied realm of marine biodiversity, however, viral and bacterial marine systematics and ecology are especially poorly known (Schuaff 1992; McCarthy and Crowder 2000). The specificity of interactions among such potential biocontrol microbes and their unicellular eukaryotic targets, then, is not well understood at the most basic autecological levels. Moreover, interactions between unicellular organisms and their hosts may vary along geographical and microhabitat gradients (e.g. Secord and Augustine 2000).

Multiple clones of the virus that attacks *Heterosigma akashiwo* differ in their efficacy at inducing cell lysis, and moreover the *Heterosigma* strains may vary greatly in their susceptibility (Nagasaki and Yamaguchi 1998). Given very high mutation rates and short generation times of microorganisms generally, biocontrol seems unlikely to be a durable solution to ecological problems associated with increases in toxic microalgae. Indeed, recent genetic evidence for viral resistance by some *H. akashiwo* clones suggests a coevolutionary history of virulence and resistance that could be accelerated with anthropogenic inoculations of bloom-ending viral particles (Tarutani et al. 2000). It seems particularly easy for such microbial control species to evolve changes in their host specificity to attack other kinds of phytoplankton, and for *Heterosigma* and other potential hosts to use existing genetic variation to evolve resistance to such biological control agents. Moreover, eukaryotic microbes that might be useful for biocontrol often engage in highly transitory mixotrophic interactions that vary depending on the environment and the presence of other species (e.g. Stoecker 1992). This combination of ecological opportunism and evolving host specificity does not imply durable damage to HAB hosts.

Biocontrol of the ctenophore Mnemiopsis leidyi *in the Black Sea*

The northwestern Atlantic ctenophore *Mnemiopsis leidyi* has achieved notoriety in the last decade as the putative cause of the collapse of the Black Sea's multispecies fishery, dealing an economic blow to several countries already in dire straits after the breakup of the former Soviet Union (Kideys 1994; Kideys et al. 2000). This fisheries decline was apparently caused indirectly by the ctenophores' competition with fish larvae for their zooplankton prey, and perhaps directly by ctenophore predation on fish larvae (although gut contents show the main prey are copepods and molluscs; Mutlu 1999). These ctenophore invaders have since spread into the Sea of Azov, the Eastern Mediterranean Sea, and the Caspian Sea (GESAMP 1997; Ivanov et al. 2000; Shiganova and Bulgakova 2000), and have become one of several 'poster children' for aquatic invasions, even appearing on the cover of a recent popular book on invasion biology (Bright 1998). At its peak, the total *Mnemiopsis* biomass reached one billion tons in the Black Sea alone (Vinogradov et al. 1989), exceeding the total world fishery catch for that year (UN Food and Agriculture Organization; data at http://apps.fao.org).

While *M. leidyi* is a voracious member of the gelatinous zooplankton community with a high trophic status, it does have a few predators of its own, some of which have been proposed for its biocontrol in its introduced range in the Black Sea. Species that may regulate *M. leidyi* populations in its northwestern Atlantic home range include the predatory scyphomedusa *Chrysaora quinquecirrha* and the parasitic larval sea anemone *Edwardsia lineata* (Purcell and Cowan 1995; Bumann and Puls 1996). The latter has been proposed as a possible biocontrol agent, though possible negative effects of such an introduction include the unknown consequences of large numbers of adult anemones on benthic communities, possible shifts of these parasites to other gelatinous zooplankton, and the fact that these anemone larvae are pests themselves in their free-swimming stage, causing 'seabather's eruption' (Bumann and Puls 1996).

The most seriously proposed biocontrol agent for *M. leidyi* in the Black Sea is a vertebrate predator, the butterfish *Peprilus triacanthus* (Harbison 2001). The fish is native to the northwestern Atlantic Ocean, a generalist predator on gelatinous zooplankton, and commercially valuable (GESAMP 1997). However, if introduced it would share many of the attributes of other vertebrate predators introduced worldwide for biocontrol, with several potential unintended consequences. Although it is not a specialist predator on *M. leidyi,* its diet is indeed largely restricted to gelatinous zooplankton (Oviatt and Kremer 1977). However, it has also been reported to prey on a range of pelagic and planktonic species, including 'tunicates, crustaceans, chaetognaths, polychaetes, ctenophores, and cnidarians' (Murawski and Waring 1979). In addition, regardless of whether it regulated ctenophore populations in the Black Sea, it could easily disperse into the Eastern Mediterranean, where it would be likely to compete with close relatives and generate market pressures to sustain its populations (R.L. Haedrich, pers. comm.). Like other pests, unwanted biocontrol agents are very difficult to eliminate once their populations reach high levels over broad geographic areas.

The challenge of controlling ctenophore numbers in the Black Sea is embedded in a complex web of environmental changes that have affected the area during the second half of the twentieth century (GESAMP 1997). For example, the biological effects of pesticide runoff, salinity changes, eutrophication, and overfishing may have synergistically created conditions favoring the success of *M. leidyi.* If some or all of these conditions were reversed, the currently large ctenophore populations might be reduced to less damaging levels. Indeed, since the ctenophore's initial population outbreak in the 1980s, its numbers have declined significantly, zooplankton have increased, and traditional anchovy and other fisheries have begun to improve (Kideys and Romanova 2001). This decline of *M. leidyi* is likely due to both bottom-up forces (Kideys et al. 2000) and top-down forces, including the (accidental?) introduction of yet another ctenophore, *Beroe ovata,* that eats *M. leidyi* (Shiganova and Bulgakova 2000).

In this case, the invader seems to be reaching a tolerable level of impact after an initially alarming one. Static host specificity of the proposed biocontrol agents is unlikely, and indeed there has been past confusion even over the taxonomic identity of the invading ctenophore itself. Many complex interacting factors have affected its population and that of native and commercially valuable species, and no clear ecological data or reasoning would have guaranteed that yet another non-indigenous species introduced at the height of the invasion would have reduced *M. leidyi* to acceptable levels with minimal side effects. In a recent review of zooplankton dynamics in the context of environmental challenges facing the Black Sea, Kideys et al. (2000) conclude that in such environmentally degraded areas, problems 'should not be exported elsewhere, but resolved within the system'. Importation of potential new problems, as well as biocontrol introductions with the potential to evolve and disperse to adjacent regions, would clearly violate this advice.

Biocontrol of the European green crab Carcinus maenas *in its introduced range*

Carcinus maenas, a hardy, omnivorous crab native to Eastern Atlantic shores, has successfully invaded temperate marine habitats in much of the rest of the world, including the east and west coasts of North America, and parts of South America, Asia, South Africa, mainland Australia, and Tasmania (Thresher 1999; Jamieson 2000). As a predator it threatens bivalve shellfish industries, and as a predator or competitor it can influence native crustacean biodiversity (McDonald et al. 2001). Although it has been 'naturalized' in some parts of its introduced range for a century or more, this is still an evolutionarily brief period (Carlton 1999). In other taxa, however, enough generations have passed in less absolute time for host shifts and host range expansions to have taken place (Secord and Kareiva 1996). In any case, there is no reason to assume a lack of important ongoing ecological, evolutionary or economic impacts of green crabs (Carlton 1999). Several proposals have recently appeared to seek, test, and deploy natural enemies of *C. maenas*, despite few experimental field data on the crab's ecological impacts (Thresher 1997; Grosholz 1999; Grosholz et al. 2000; McDonald et al. 2001).

A range of natural enemies of *C. maenas* has been suggested for its biological control, in both scientific and popular media (Blair 1998). These include viruses, dinoflagellates, ciliates, nemerteans, and rhizocephalans (parasitic castrating barnacles; Goggin 1997; Thresher 1996, 1997). The first three are either insufficiently host specific, insufficiently virulent, or, given their status as microorganisms, too likely to evolve to exploit new hosts (Secord and Kareiva 1996; Goggin 1997). The metazoans have thus become the focus of biocontrol research. Nemerteans in the genus *Carcinonemertes* are symbiotic egg predators on crabs, including *C. maenas*. However, they appear to be insufficiently host specific for biological control, attacking a range of native crabs as well (Wickham and Roe 1987; Torchin et al. 1996; Goggin 1997).

More seriously considered for green crab biocontrol is a remarkably intimate parasitic symbiont, the castrating barnacle *Sacculina carcini*. By way of a highly complex life cycle, it invades the bodies of male and female crabs, castrating both sexes and behaviorally and morphologically feminizing the males (Lafferty and Kuris 1996; Thresher et al. 2000). An infected crab in essence becomes a 'parasite genotype with a crab phenotype' (Thresher et al. 2000). Risks associated with the introduction of such parasites include the ecological or evolutionary transfer of parasite populations onto native and/or commercially valuable crab or shrimp hosts. To evaluate this risk, detailed experimental work was undertaken to assess the likelihood of host shifts or host range expansions onto native Australian and Californian crabs.

Thresher et al. (2000) conducted terrestrial-insect-inspired host specificity tests evaluating the mechanisms by which the European *S. carcini* and a native Australian rhizocephalan, *Heterosaccus lunatus*, choose crabs to invade. *S. carcini* was capable of settling on many species of hosts in at least two families, though it does not complete its development to the same degree on all of them. Nevertheless, the genetic variability exists for an unacceptably broad range of target hosts. Given that rhizocephalans are morphologically depauperate and therefore traditional systematics (Schauff 1992) yields suspect species identifications, Murphy and Goggin (2000) compared 3′ region SSU and ITS1 rRNA nucleotide sequences from putative *S. carcini* taken from a variety of host species in several locations. Adding to the results of Thresher et al. (2000), they found that the same nominal species infects three crab species in two genera. Thus, *S. carcini* for biocontrol of European green crabs in their introduced Australian range may pose a risk to native or marketable crabs there. Similar host specificity experiments in California, also part of the introduced range of *C. maenas*, found settlement and development of *S. carcini* interna on four species of native crabs, including the economically important Dungeness crab *Cancer magister* (Goddard et al. 2001). Moreover, *S. carcini* killed these species before maturing in them (Goddard et al. 2001), including those in two families not known to be infected by *S. carcini* in nature.

A final caution against using these parasitic barnacles as biocontrol agents comes from broader taxonomic studies of their host ranges. Høeg and Lutzen (1985) found that the host range of the family Sacculinidae as a whole is quite broad, including many species in *seven families* of brachyuran crabs, including the commercially important Cancridae and Majidae. This familial adaptive radiation, together with the degree of host specificity in the individual species discussed above, suggest that HS and HRE are likely if we introduce these parasites to new ocean basins with high crab species richness.

Biocontrol of the predatory starfish Asterias amurensis

The starfish *Asterias amurensis* is native to Asian coastal zones, but has recently been introduced via ballast water to Tasmania in the temperate south Pacific. Massive settlement and metamorphosis by its planktotrophic larvae have resulted in adult populations as great as 30 million individuals in the Derwent River and adjacent estuaries in southern Tasmania (Goggin 1998a, b). Since its initial discovery and outbreak in Tasmanian waters, *A. amurensis* has spread to the Australian mainland in the state of Victoria (Garnham 1998). A voracious omnivore, it affects native organisms at the population and community levels (Ross and Johnson 1998), and has the potential seriously to disrupt aquaculture (Whayman 1998). Indeed, its appearance in Australian waters motivated the development of an impressive infrastructure for research and policy response to marine invasive species in that country (Turner 1998).

Following guidelines developed for identifying potential biocontrol agents for marine pest species, Kuris et al. (1996) listed four candidate species that might be used against *A. amurensis*. These included two parasitic crustaceans (a copepod and an ascothoracican), one parasitic gastropod, and a spermiphagous ciliate. Three of these species affect seastar reproduction, and one causes somatic necrosis. Of these candidate species, a preliminary analysis in the native range of *A. amurensis* suggested that the ciliate *Orchitophyra stellarum* presented the most promise for biocontrol, given its strongly negative effects on male reproduction (Kuris et al. 1996). However, field data are lacking on whether this ciliate can actually regulate seastar populations (Goggin and Bouland 1997).

Furthermore, there is strong evidence that this species is not host specific. Byrne et al. (1997) explicitly caution against its use for biocontrol because it is capable of infecting several genera of starfishes. Of particular concern is the observation that *O. stellarum* causes castration and death in the northeast Pacific starfish *Pisaster ochraceus* (Leighton et al. 1991), a well-known keystone species on rocky shores (Paine 1966). While this parasite is apparently already on the Pacific coast of Canada, it is cryptogenic (see Leighton et al. 1991), leaving the risk of additional trans-Pacific introductions in ballast water from Australia to the west coast of North America. Further, the possibility remains that this parasite could expand its host range to include native – and ecologically important – starfishes of the South Pacific.

Biocontrol of the green alga Caulerpa taxifolia

The toxic, siphonaceous green macroalga *Caulerpa taxifolia* has joined the ranks of the world's best known marine plant invaders. The 'aquarium strain' of this alga was developed from tropical origins as a hardy, cold-tolerant aquarium plant used in many parts of the world. First detected outside captivity in a square meter patch beneath the Oceanographic Institute in Monaco in 1984, the alga has now spread to cover thousands of hectares of substrata along the shores of six Mediterranean countries from Croatia to Tunisia (Meinesz 1999; Thibaut and Meinesz 2001). More recently, it has been discovered on the coasts of California and Australia, where eradication efforts are underway (Dalton 2000; Jousson et al. 2000). *C. taxifolia* spreads rapidly by vegetative growth and fragmentation, complicating mechanical control efforts and enhancing inadvertent dispersal by boats and anchors.

Members of several taxa of oligophagous sacoglossan sea slugs attack *Caulerpa* and related algae, either through kleptoplasty (the 'theft' of chloroplasts) or through outright herbivory (Jensen 1997). As the invasion of *C. taxifolia* has progressed in the Mediterranean, there have been increasing calls for the use of such slugs in the biological control of these toxic algae (e.g. Meinesz 1999; Coquillard et al. 2000; Thibaut and Meinesz 2000; Thibaut et al. 2001). Meinesz (1999) states that due to slow public action and governmental ineptitude in the early years of the invasion, it is no longer possible to control *C. taxifolia* by any method other than biocontrol. Thibaut and Meinesz (2000) evaluated two native Mediterranean sacoglossans, *Oxynoe olivacea* and *Lobiger serradifalci,* for their efficacy in controlling invasive *C. taxifolia*. They concluded that while these two species are known to consume *C. taxifolia,* their low feeding rates and planktotrophic larval development make them unsuitable for biocontrol by 'enhancement' of native natural enemies. Moreover, the latter sacoglossan, in the process of feeding, apparently fragments *C. taxifolia*, thus enhancing its clonal dispersal ability (Zuljevic et al. 2001).

Thibaut and Meinesz (2000, 2001) recommend, instead, the importation of *Elysia subornata* and *Oxynoe azuropunctata*, faster-feeding, locally dispersing sacoglossans from tropical northwest Atlantic regions. They suggest that these species are host-specific feeders on *Caulerpa* spp., and that as tropical animals, they will not survive Mediterranean winters. The latter assumption is suspect, however, given that *C. taxifolia*, thriving in the Mediterranean currently, also had its origins in the tropics and was not expected to thrive in temperate regions such as the Mediterranean and California. Simulation models of classical biocontrol using *E. subornata* for the Mediterranean indicate that low water temperatures could have a strong negative effect on the slug's biocontrol potential, and that slugs would need to occur at relatively high densities (4/m^2) for any significant slowing of algal growth (Coquillard et al. 2000). Thibaut and Meinesz (2001) suggest that *E. subornata,* if introduced, would not compete with native sacoglossans until 'all *Caulerpa* in the Mediterranean have been completely eradicated'. However, no spatial or demographic data support this conclusion, especially if the introduced slugs (whose general biology is not well known) expand their host affinities to include native *Caulerpa* species.

Further, Thibaut and Meinesz (2001) suggest that a public awareness campaign could prevent dispersal of biocontrol slugs from their initial release zones; the mechanism by which this would happen, however, is unclear. Given that the slugs are cryptic and very small (typically a few mm in length), and the fact that increased public awareness has not appreciably halted the spread of the much larger and more conspicuous *Caulerpa* itself, this outcome seems unlikely.

Finally, extensive ecological and evolutionary work on host preferences of sacoglossans strongly suggests that any 'guarantee' of durable host specificity in these slugs should be viewed with great skepticism. While it has long been assumed that sacoglossans have highly specific feeding preferences, evidence has been accumulating for many years that a number of species have relatively broad diets (Williams and Walker 1999) and frequently switch their favored host species (Clark and Busacca 1978; Jensen 1997). For example, a congener of the proposed classical biocontrol species, *Elysia viridis,* is capable of learning through 'ingestive conditioning' to switch its dietary preferences from one green algal genus to another (Jensen 1989). Jensen (1994) documents additional behavioral variations subject to selection that are important in the evolution of sacoglossan diets; these include food handling time and methods and animals' response to spatial arrangement of prey. Further, she showed high variation in specificity among sacoglossan species, suggesting that feeding specificity is subject to rapid evolutionary change, especially among temperate species where seasonality of food places strong selection pressures on slugs (Jensen 1994, 1997). Relatively high levels of intraspecific genetic variation in sacoglossans may provide a mechanism (especially in the temperate waters of the Mediterranean) for such rapid shifts in host preference in ecological and evolutionary time (Theisen and Jensen 1991; Secord and Kareiva 1996).

Recent experimental evidence indicates that host shifts and host range expansions *among native and introduced algal hosts by native sacoglossan slugs* are frequent occurrences in historical time (Trowbridge and Todd 2001; C. Trowbridge, unpub. ms). As they have in other regions, native sacoglossans in the Mediterranean have expanded their host ranges to include the invasive *C. taxifolia*. It is reasonable, then, to expect that this rapid evolution in host preference will progress to increased feeding efficiencies on this novel host. As suggested by Trowbridge (1999) for control of *Codium fragile,* greater efforts should be made in artificial selection and augmentative biocontrol using native natural enemies (even those with planktonic larvae) before risking the introduction of yet more exotic species in the Mediterranean.

The example of *C. taxifolia* in the Mediterranean presents a powerful lesson in lost opportunities for eradication at very early stages of an invasion (Meinesz 1999). Eradication of infestations of potential pests when populations are small, by methods that (unlike classical biocontrol introductions) are not irreversible, is an overlooked early response that may be appropriate for many kinds of invasions (Dahlsten et al. 1989; Bax 1999; Thresher 1999; Culver and Kuris 2000; Simberloff 2001).

Biocontrol of the saltmarsh cordgrass Spartina alterniflora *with the planthopper* Prokelesia marginata

Saltmarsh cordgrasses in the genus *Spartina* are especially successful invaders in mudflat habitats that previously lacked emergent salt marsh macrophytes (Daehler and Strong 1996). This negatively affects diverse intertidal and subtidal communities of infaunal crustaceans, polychaetes and bivalves, including commercially important native crabs, native and non-indigenous cultivated oysters, and invertebrate prey for migrating seabirds. *Spartina* also modifies sediment deposition regimes, raising substrate and essentially transforming intertidal mudflats into semi-terrestrial grasslands (Daehler and Strong 1996). A major political and scientific response has occurred in recent years in Willapa Bay, a large estuary adjoining the outer coast of Washington State. Massive and controversial chemical and mechanical control efforts have been deemed ineffective, leading to calls to introduce a biocontrol agent as part of an IPM program for *S. alterniflora*. The biocontrol species currently undergoing field releases is the planthopper *Prokelesia marginata*, a member of a group of homopteran insects whose ecology, evolution, and management is relatively well understood compared to most marine taxa (e.g. see Denno and Perfect 1994 for a recent 800-page volume on the taxon).

While not risk-free, on the face of it *Spartina* biocontrol may appear to differ in several important respects from the foregoing examples. First, in many ways it is not truly marine biocontrol but rather 'terrestrial biocontrol by the sea' (A.M. Kuris, pers. comm.). It involves two taxa – insects and flowering plants – that are perhaps the best understood coevolved clades of natural enemies (Ehrlich and Raven 1964; Thompson 1994; Jolivet 1998). The biocontrol program involves extensive community involvement, a high level of local support, and participation by biologists who are expert in the ecological and evolutionary dynamics of the partner taxa *and who have historically been outspoken biocontrol skeptics* (Cory and Myers 2000). Genetic data suggest that *S. alterniflora* in Willapa Bay is unusually susceptible to predation by *P. marginata* (Daehler and Strong 1995, 1997). The biocontrol species is not being brought from another ocean basin or another continent, but rather from California, farther south along the same coast, where it (and another *Spartina* species) is native; no native *Spartina* species exists in Pacific Northwest estuaries (Daehler and Strong 1996). Finally, detailed and well-funded specificity testing has been performed using standard methods of experimental ecology (CPRC 2000); the semi-terrestriality of the habitats makes cage experiments possible in the field, in contrast with several of the other cases of marine biocontrol discussed here.

Despite the above features that would appear to make it more promising than the other case studies examined here, it is prudent to retain a high level of caution regarding this biocontrol introduction. The mechanism by which the planthopper damages *Spartina* is not fully understood, and unexplained roles for bacteria in the insects' sucking mouthparts need further study (Wu et al. 1999; S. Hacker, pers. comm.). Native *Spartina* species occur east of the Cascade Mountains in Washington State, far from coastal estuaries but certainly within range of anthropogenic dispersal mechanisms. And it is reasonable to expect that if the insects are successful at controlling *Spartina,* they will experience exceptionally strong selection pressure to utilize new hosts. Ultimately, the environmental risks of any biocontrol program must balance the tremendous uncertainties of additional species introductions against the long-term damage done by the invader itself.

Is biological control in the sea different? Terrestrial *versus* marine biological control

Marine biocontrol contrasts with its terrestrial counterpart in a number of important respects. First, it is much newer. There is vastly less experience and many fewer foundational data to draw upon in making informed decisions about risk and efficacy. Tens of thousands of citations appear in the empirical and theoretical literature on terrestrial biocontrol over more than a century, as opposed to at most a few dozen papers and reports, nearly all cited herein, for marine biocontrol (Huffaker et al. 1971; papers in Hawkins and Cornell 1999; Perkins and Garcia 1999). Second, hundreds of successful and failed examples of biocontrol exist on land, whereas in the sea there are only a handful of proposals with only one at an advanced stage of implementation (*Prokelesia* field tests to control *Spartina* in Willapa Bay, Washington, USA). Third, marine systems are much more diverse at higher taxonomic levels, containing almost all animal and plant phyla, many of which are especially poorly known. A clear understanding of systematics is critical in assessing host specificity and non-target effects (Schauff 1992). Indeed, marine taxonomy has never had the large economic engine of agriculture to help it mature to the point that terrestrial plant and insect systematic research have. Terrestrial life represents only a small subset of higher-order biodiversity, and the ecologically dominant taxa there are exceptionally (even if imperfectly) well understood in their trophic and coevolutionary relationships (Ehrlich and Raven 1964; Schauff 1992; Thompson 1994; Jolivet 1998). In short, we simply know more about the genetic mechanisms and environmental contexts underlying host shifts and host range expansions in weevils and grasses than we do in jellies or algae. Finally, marine taxa such as starfishes and ctenophores have very different life histories and morphologies (e.g. planktonic feeding larvae, noncephalized body plans) and biomechanical regimes (e.g. dispersal and host-seeking in a high-density fluid medium) than do their terrestrial counterparts (Strathmann 1990).

Potentially harmful bioinvasions can be addressed through a series of variably risky strategies: (1) prevention of invasions in the first place (minor economic costs but otherwise very low risk); (2) eradication of invasive populations when they are small and easily detected and eliminated (moderate risk due to localized chemical or mechanical control efforts); (3) control of established populations with mechanical methods or limited use of chemicals (moderate risk depending on spatial and temporal extent of application); (4) augmentative biocontrol with native species (moderate risk since no new introductions, but community dynamics are potentially altered); or (5) classical or neoclassical biological control (highest risk due to irreversibility and ecological and evolutionary uncertainty). Of course, most IPM programs use combinations of these approaches.

I recommend here, as others have, that biological control in the sea should be considered a last resort, and used only when the risks posed by a particular invasion warrant it (Bax et al. 2001; Hemmelgarn et al. 2002). For one, the level of uncertainty in marine systems is simply far greater than it is in terrestrial systems. Because of this uncertainty, biocontrol might even be inappropriate for marine invasions initially considered catastrophic. This need not mean abandonment of all hope. Indeed, some invasive populations seem to decline as dramatically as they initially took off, perhaps reducing the apparent need for biocontrol introductions (e.g. *Mnemiopsis* in the Black Sea and the Asian clam *Potamocorbula amurensis* in San Francisco Bay; Mutlu 1999; P.V. Scott, pers. comm.).

Alternative strategies for dealing with marine invasions are in various stages of consideration in different parts of the world. Approaches to prevention vary with the particular anthropogenic dispersal vectors involved, and include appropriate public education and outreach programs, ballast water exchange and treatment options, and regulation of the aquaculture industry. Still, we have plenty of work left to do, and money spent on prevention of invasions will preclude new, damaging invasions and the subsequent need for risky control efforts (Bax et al. 2001; Hemmelgarn et al. 2002). Nevertheless, no prevention program is perfect and species will continue to invade new habitats. In such cases, early detection and eradication of the invasion will preclude the need to discuss biological control later once it has spread. Monitoring programs, while considered by most ecologists to be boring from a scientific point of view, may be very good public policy if they involve taxonomic specialists, ecological generalists, and the general public in identifying and rapidly responding to new invasions (Mills et al. 1999; Cohen et al. 2001). This post-detection rapid response may include limited chemical control or mechanical control such as shading of algal species, chlorine treatment of enclosed marinas, or physical smashing of individual snail hosts of an invasive parasite (Bax 1999; Culver and Kuris 2000; Dalton 2000). Of course, the environmental impacts on other species of such activities must be seriously considered, and the least damaging action chosen; even occasional chemical control, though, does not have the evolutionary uncertainty and permanence of classical biocontrol.

Biological methods other than classical biocontrol should also be more fully evaluated for their efficacy. For example, the use of triploid or otherwise sterile organisms, artificial selection and mass rearing technologies for native enemies, and 'population inviability analysis' to attack pest populations at their most vulnerable stages all have strong potential for use in marine pest control (Grosholz 1999). Augmentative biocontrol using native species should also be given high priority in marine habitats (it has been underutilized in terrestrial biocontrol) when prevention, early detection and eradication fail (Chang and Kareiva 1999). Many examples considered here clearly illustrate the frequency with which HS and HRE evolve in both marine and terrestrial taxa. Instead of risking attacks on non-target native organisms by introduced biocontrol agents, whenever possible why not encourage native 'enemies' to attack introduced marine hosts that need to be controlled? As documented above, such host shifts and host range expansions have frequently occurred eventually anyway.

Biologists and policy makers need to harness the full arsenal of modern marine biology and ecology in the prevention and control of harmful marine invaders. This presents us with many more options than are commonly recognized, and surely we can do more than simply introducing additional non-indigenous species to control pests without exhausting the alternatives. Terrestrial biological control has worked spectacularly well in some cases, but it has also caused unforeseen problems. Life in the sea is more diverse and fundamentally different than life on land and we should be cautious about exporting the assumptions and lessons of agricultural biocontrol without a great deal of thought and even more data.

Acknowledgements

I thank John Chapman, Andy Cohen, Jeff Goddard, Sally Hacker, Peter Kareiva, Armand Kuris, Alexandre Meinesz, Gisele Muller-Parker, Bob Paine, Cynthia Trowbridge, Miranda Wecker, Marjorie Wonham, and many other colleagues and students for stimulating discussions on the ecology of marine symbioses and/or appropriate criteria for marine biocontrol. Amy Adams, Lowell Adams, John Banks, Jim Carlton, Richard Haedrich, Kristin Hemmelgarn, and Claudia Mills provided critical references or comments that greatly improved this paper; the detailed criticisms of three anonymous reviewers also honed my thinking on the prospects and risks of marine biocontrol. Marilyn O'Leary and Judy Pederson provided the opportunity to further explore the topic in a panel discussion at the Second International Conference on Marine Bioinvasions in New Orleans in April 2001. I thank them, panelists Armand Kuris, Alexandre Meinesz and Marjorie Wonham, and audience participants for their critical thinking on a highly complex issue. I am grateful for financial support from the National and Washington State Sea Grant programs (NOAA) and from the University of Washington, Tacoma Founders' Endowment Fund. Cynthia Catton, Kristin Hemmelgarn, Tonya Kauhi, and Tim Strickler relentlessly helped track down key references. This is contribution #1003 of the Center for the Study of Invertebrates and Society at the University of Washington, Tacoma.

References

Balciunas JK, Burrows DW and Purcell MF (1996) Comparison of the physiological and realized host-ranges of a biological control agent from Australia for the control of the aquatic weed, *Hydrilla verticillata*. Biological Control 7: 148–158

Barbosa P (1998) Conservation Biological Control. Academic Press, San Diego, California, 396 pp

Barratt BIP, Evans AA, Ferguson CM, Barker GM, McNeill MR and Phillips CB (1997) Laboratory nontarget host range of the introduced parasitoids *Microctonus aethiopoides* and *M. hyperodae* (Hymenoptera: Braconidae) compared with field parasitism in New Zealand. Environmental Entomology 26: 694–702

Barratt BIP, Ferguson CM and Evans AA (2001) Non-target effects of introduced biological control agents and some implications for New Zealand. In: Lockwood JA, Howarth FG and Purcell MF (eds) Balancing Nature: Assessing the Impact of Importing Non-native Biological Control Agents (An International Perspective), pp 41–53. Entomological Society of America, Lanham, Maryland

Bax NJ (1999) Eradicating a dreissenid from Australia. Dreissena! 10: 1–5

Bax N, Carlton JT, Mathews-Amos A, Haedrich RL, Howarth FG, Purcell JE, Rieser A and Gray A (2001) The control of biological invasions in the world's oceans. Conservation Biology 15: 1234–1246

Bigger DS and Marvier MA (1998) How different would a world without herbivory be? A search for generality in ecology. Integrative Biology 1: 60–67

Blair T (1998) Lateral thought on the sea bed: the fix for a European pest may be its European foe. Time, 30 March 1998 (issue 13)

Blossey B (1995) Host specificity screening of insect biological weed control agents as part of an environmental risk assessment. In: Hokkanen HMT and Lynch JN (eds) Biological Control: Benefits and Risks. Cambridge University Press, Cambridge

Blossey B, Schroeder D, Hight SD and Malecki RA (1994a) Host specificity and environmental impact of the weevil *Hylobius transversovittatus*, a biological control agent of purple loosestrife (*Lythrum salicaria*). Weed Science 42: 128–133

Blossey B, Schroeder D, Hight SD and Malecki RA (1994b) Host specificity and environmental impact of two leaf beetles (*Galerucella calmariensis* and *G. pusilla*) for biological control of purple loosestrife (*Lythrum salicaria*). Weed Science 42: 134–140

Boettner GH, Elkinton JS and Boettner CJ (2000) Effects of a biological control introduction on three nontarget native species of saturniid moths. Conservation Biology 14: 1798–1806

Bratbak G, Egge JK and Heldal M (1993) Viral mortality of the marine alga *Emiliania huxleyi* (Haptophyceae) and termination of algal blooms. Marine Ecology Progress Series 93: 39–48

Bright C (1998) Life out of Bounds: Bioinvasion in a Borderless World. W.W. Norton, New York, 287 pp

Brooks DR and McLennan DA (1993) Parascript: Parasites and the Language of Evolution. Smithsonian Institution Press, Washington DC, 429 pp

Brussaard CPD, Kempers RS, Kop AJ, Riegman R and Heldal M (1996) Virus-like particles in a summer bloom of *Emiliania huxleyi* in the North Sea. Aquatic Microbial Ecology 10: 105–113

Bumann D and Puls G (1996) Infestation with larvae of the sea anemone *Edwardsia lineata* affects nutrition and growth of the ctenophore *Mnemiopsis leidyi*. Parasitology 113: 123–128

Byrne M, Cerra A, Nishigaki T and Hoshi M (1997) Infestation of the testes of the Japanese sea star *Asterias amurensis* by the ciliate *Orchitophyra stellarum*: a caution against the use of this ciliate for biological control. Diseases of Aquatic Organisms 28: 235–239

Carlton JT (1999) Quo vadimus exotica oceanica? Marine bioinvasion ecology in the twenty-first century. In: Pederson J (ed) Marine Bioinvasions: Proceedings of the First National Conference, January 24–27, 1999, pp 6–23. Massachusetts Institute of Technology, Cambridge, Massachusetts

Carroll SP and Boyd C (1992) Host race radiation in the soapberry bug: natural history with the history. Evolution 46: 1052–1069

Chang GC and Kareiva P (1999) The case for indigenous generalists in biological control. In: Hawkins BA and Cornell HV (eds) Theoretical Approaches to Biological Control, pp 103–115. Cambridge University Press, Cambridge

Clark KB and Busacca M (1978) Feeding specificity and chloroplast retention in four tropical ascoglossa, with a discussion of the extent of chloroplast symbiosis and the evolution of the order. Journal of Molluscan Studies 44: 272–282

Clarke B, Murray J and Johnson MS (1984) The extinction of endemic species by a program of biological control. Pacific Science 38: 97–104

Cohen AN, Berry HD, Mills CE, Milne D, Britton-Simmons K, Wonham MJ, Secord DL, Barkas JA, Bingham B, Bookheim BE, Byers JE, Chapman JW, Cordell JR, Dumbauld B, Fukuyama A, Harris LH, Kohn AJ, Li K, Mumford Jr TF, Radashevsky V, Sewell AT and Welch K (2001) Washington State Exotics Expedition 2000: a rapid assessment survey of exotic species in the shallow waters of Elliott Bay, Totten and Eld Inlets, and Willapa Bay. Nearshore Habitat Program, Washington State Department of Natural Resources, Olympia, Washington, 47 pp

Coquillard P, Thibaut T, Hill DRC, Gueugnot J, Mazel C and Coquillard Y (2000) Simulation of the mollusc Ascoglossa *Elysia subornata* population dynamics: application to the potential biocontrol of *Caulerpa taxifolia* growth in the Mediterranean Sea. Ecological Modelling 135: 1–16

Cory JS and Myers JH (2000) Direct and indirect ecological effects of biological control. Trends in Ecology and Evolution 15: 137–139

CPRC (2000) UC-Davis team awarded $3.8 million NSF grant to study invasive *Spartina*. Columbia Pacific Resources Center (Naselle, WA) *Spartina* Biocontrol News 7: 4

Crawley MJ (ed) (1992) Natural Enemies: the Population Biology of Predators, Parasites and Diseases. Blackwell Scientific Publishers, London, 576 pp

Culver CS and Kuris AM (2000) The apparent eradication of a locally established introduced marine pest. Biological Invasions 2: 245–253

Daehler CC and Strong DR (1995) Impact of high herbivore densities on introduced smooth cordgrass, *Spartina alterniflora*, invading San Francisco Bay, California. Estuaries 18: 409–417

Daehler CC and Strong DR (1996) Status, prediction and prevention of introduced cordgrass *Spartina* spp. invasions in Pacific estuaries, USA. Biological Conservation 78: 51–58

Daehler CC and Strong DR (1997) Reduced herbivore resistance in introduced smooth cordgrass (*Spartina alterniflora*) after a century of herbivore-free growth. Oecologia 110: 99–108

Dahlsten DL, Garcia R and Lorraine H (eds) (1989) Eradication of Exotic Pests: Analysis with Case Histories. Yale University Press, New Haven, Connecticut, 296 pp

Dalton R (2000) Researchers criticize response to killer algae. Nature 406: 447

Dennill GB, Donnelly D and Chown SL (1993) Expansion of host range of a biocontrol agent *Trichilogaster acaciaelongifoliae* (Pteromalidae) released against the weed *Acacia longifolia* in South Africa. Agriculture Ecosystems and the Environment 43: 1–10

Denno RF and Perfect TJ (eds) (1994) Planthoppers: Their Ecology and Management. Chapman & Hall, New York

Dent D (1995) Integrated Pest Management. Chapman & Hall, London

Doutt RL (1958) Vice, virtue and the vedalia. Bulletin of the Entomological Society of America 4: 119–123

Doutt RL and Smith RF (1971) The pesticide syndrome – diagnosis and suggested prophylaxis. In: Huffaker CB (ed) Biological Control, pp 3–15. Plenum Press, New York

Ehler LE (1998) Invasion biology and biological control. Biological Control 13: 127–133

Ehrlich PR and Raven PH (1964) Butterflies and plants: a study in coevolution. Evolution 18: 586–608

Elliott N, Kieckhefer R and Kauffman W (1996) Effects of an invading coccinellid on native coccinellids in an agricultural landscape. Oecologia 105: 537–544

Follett PA and Duan JJ (eds) (2000) Nontarget Effects of Biological Control. Kluwer Academic Publishers, Dordrecht, The Netherlands

Garnham J (1998) Distribution and impact of *Asterias amurensis* in Victoria. In: Goggin CL (ed) Proceedings of a Meeting on the Biology and Management of the Introduced Seastar *Asterias amurensis* in Australian Waters, pp 18–21. Technical report no 15, Centre for Research on Introduced Marine Pests (CSIRO Division of Fisheries), Hobart, Australia

Gause GF (1969) The Struggle for Existence. Hafner, New York, 163 pp

GESAMP [IMO/FAO/UNESCO–IOC/WMO/WHO/IAEA/UN/UNEP Joint Group of Experts on the Scientific Aspects of Marine Environmental Protection] (1997) Opportunistic settlers and the problem of the ctenophore *Mnemiopsis leidyi* invasion in the Black Sea. Rep. Stud. GESAMP, 58, 84 pp

Goddard JHR, Torchin ME, Lafferty KD and Kuris AM (2001) Experimental infection of native California crabs by *Sacculina carcini*, a potential biocontrol agent of introduced European green crabs. Abstract Booklet, Second International Conference on Marine Bioinvasions, pp 54–55, New Orleans, 9–11 April 2001. Massachusetts Institute of Technology Sea Grant Program, Cambridge, Massachusetts

Goggin L (1997) Parasites (excluding *Sacculina*) which could regulate populations of the European green crab *Carcinus maenas*. In: Thresher RE (ed) Proceedings of the First International Workshop on the Demography, Impacts and Management of Introduced Populations of the European Crab, *Carcinus maenas*, pp 87–91. Technical report no 11, Centre for Research on Introduced Marine Pests (CSIRO Division of Fisheries), Hobart, Australia

Goggin CL and Bouland C (1997) The ciliate *Orchitophyra* cf. *stellarum* and other parasites and commensals of the northern pacific seastar *Asterias amurensis* from Japan. International Journal for Parasitology 27: 1415–1418

Goggin CL (ed) (1998a) Proceedings of a Meeting on the Biology and Management of the Introduced Seastar *Asterias amurensis* in Australian Waters. Technical report no 15, Centre for Research on Introduced Marine Pests (CSIRO Division of Fisheries), Hobart, Australia

Goggin CL (1998b) Options for biological control of *Asterias amurensis*. In: Goggin CL (ed) Proceedings of a Meeting on the Biology and Management of the Introduced Seastar *Asterias amurensis* in Australian Waters, pp 53–58. Technical report no 15, Centre for Research on Introduced Marine Pests (CSIRO Division of Fisheries), Hobart, Australia

Grosholz E (1999) Ecological and evolutionary consequences of invasions: addenda to the agenda. In: Pederson J (ed) Marine Bioinvasions: Proceedings of the First National Conference, January 24–27, 1999, pp 147–153. Massachusetts Institute of Technology, Cambridge, Massachusetts

Grosholz ED, Ruiz GM, Dean CD, Shirley KA, Maron JL and Connors PG (2000) The impacts of a nonindigenous marine predator on multiple trophic levels. Ecology 81: 1206–1224

Habeck DH (1983) The potential of *Parapoynx stratiotata* L. as a biological control agent for Eurasian watermilfoil. Journal of Aquatic Plant Management 21: 26–29

Hager HA and McCoy KD (1998) The implications of accepting untested hypotheses: a review of the effects of purple loosestrife (*Lythrum salicaria*) in North America. Biodiversity and Conservation 7: 1069–1079

Harbison R (2001) The potential of the butterfish, *Peprilus triacanthus*, as an agent of biological control in the Black and Caspian Seas. Presentation at the First International Meeting, The invasion of the Caspian Sea by the comb jelly *Mnemiopsis* – problems, perspectives, need for action. Baku, Azerbaijan, April 2001 (see http://www.caspianenvironment.org/mnemiopsis/mnem_attachl9.htm)

Hawkins BA and Cornell HV (eds) (1999) Theoretical Approaches to Biological Control. Cambridge University Press, Cambridge

Hawkins BA and Marino PC (1997) The colonization of native phytophagous insects in North America by exotic parasitoids. Oecologia 112: 566–571

Hemmelgarn K, Secord D, Rupp J and Heimowitz P (2002) Reducing the impact: using public aquarium exhibits for aquatic nonindigenous species education and outreach. Aquatic Invaders 13(1): 1–7

Henneman ML and Memmott J (2001) Infiltration of a Hawaiian community by introduced biological control agents. Science 293: 1314–1316

Høeg J and Lutzen J (1985) Crustacea: Rhizocephala. Marine Invertebrates of Scandinavia, No 6. Norwegian University Press, Oslo, 92 pp

Hokkanen H and Pimentel D (1984) New approach for selecting biological control agents. The Canadian Entomologist 116: 1109–1121

Holt RD and Hochberg ME (2001) Indirect interactions, community modules and biological control: A theoretical perspective. In: Wajnberg E, Scott JK and Quimby PC (eds) Evaluating Indirect Ecological Effects of Biological Control, pp 13–37. CABI Publishing, New York

Howarth FG (1991) Environmental impacts of classical biological control. Annual Review of Entomology 36: 485–509

Huffaker CB, Messenger PS and DeBach P (1971) The natural enemy component in natural control and the theory of biological control. In: Huffaker CB (ed) Biological Control, pp 16–67. Plenum Press, New York

Ivanov VP, Kamakin AM, Ushivtzev VB, Shiganova T, Zhukova O, Aladin N, Wilson SI, Harbison GR and Dumont HJ (2000) Invasion of the Caspian Sea by the comb jellyfish *Mnemiopsis leidyi* (Ctenophora). Biological Invasions 2: 255–258

Jamieson GS (2000) European green crab, *Carcinus maenas*, introductions in North America: differences between the Atlantic and Pacific experiences. In: 10th International Aquatic Nuisance Species and Zebra Mussel Conference, pp 307–316, Toronto, Canada, February 13–17, 2000. Conference Administrator, Pembroke, Canada

Jensen KR (1989) Learning as a factor in diet selection by *Elysia viridis* (Montagu) (Opisthobranchia). Journal of Molluscan Studies 55: 79–88

Jensen KR (1994) Behavioral adaptations and diet specificity of sacoglossan opisthobranchs. Ethology Ecology and Evolution 6: 87–101

Jensen KR (1997) Evolution of the Sacoglossa (Mollusca, Opisthobranchia) and the ecological associations with their food plants. Evolutionary Ecology 11: 301–335

Jolivet P (1998) Interrelationship Between Insects and Plants. CRC Press, Boca Raton, Florida, 309 pp

Jousson O, Pawlowski J, Zaninetti L, Zechman FW, Dini F, DiGujiseppe G, Woodfield R, Millar A and Meinesz A (2000) Invasive alga reaches California. Nature 408: 157–158

Kideys AE (1994) Recent dramatic changes in the black sea ecosystem: the reason for the sharp decline in Turkish anchovy fisheries. Journal of Marine Systems 5: 171–181

Kideys AE and Romanova Z (2001) Distribution of gelatinous macrozooplankton in the southern Black Sea during 1996–1999. Marine Biology 139: 535–547

Kideys AE, Kovalev AV, Shulman G, Gordina A and Bingel F (2000) A review of zooplankton investigations of the Black Sea over the last decade. Journal of Marine Systems 24: 355–371

Kim MC, Yoshinaga I, Imai I, Nagasaki K, Itakura S and Ishida Y (1998) A close relationship between algicidal bacteria and termination of *Heterosigma akashiwo* (Raphidophyceae) blooms in Hiroshima Bay, Japan. Marine Ecology Progress Series 170: 25–32

Kuris AM, Lafferty KD and Grygier MJ (1996) Detection and preliminary evaluation of natural enemies for possible biological control of the northern pacific seastar, *Asterias amurensis*. Technical report no 3, Centre for Research on Introduced Marine Pests, 12 pp

Lafferty KD and Kuris AM (1996) Biological control of marine pests. Ecology 77: 1989–2000

Leighton BJ, Boom JDG, Bouland C, Hartwick EB and Smith MJ (1991) Castration and mortality in *Pisaster ochraceus* parasitized by *Orchitophrya stellarum* (*Ciliophora*). Diseases of Aquatic Organisms 10: 71–73

Louda SM, Kendall D, Connor J and Simberloff D (1997) Ecological effects of an insect introduced for the biological control of weeds. Science 277: 1088–1090

McCarthy HP and Crowder LB (2000) An overlooked scale of global transport: phytoplankton species richness in ships' ballast water. Biological Invasions 2: 321–322

McDonald PS, Jensen GC and Armstrong DA (2001) The competitive and predatory impacts of the nonindigenous crab *Carcinus maenas* (L.) on early benthic phase Dungeness crab *Cancer magister* Dana. Journal of Experimental Marine Biology and Ecology 258: 39–54

Meinesz A (1999) Killer Algae. University of Chicago Press, Chicago, 359 pp

Milligan KLD and Cosper EM (1994) Isolation of virus capable of lysing the brown tide microalga, *Aureococcus anophagefferens*. Science 266: 805–807

Mills CE, Cohen AN, Berry HK, Wonham MJ, Bingham B, Bookheim B, Carlton JT, Chapman JW, Cordell J, Harris LH, Klinger T, Kohn AJ, Lambert C, Lambert G, Li K, Secord DL and Toft J (1999) The 1998 Puget Sound Expedition: A shallow-water rapid assessment survey for nonindigenous species, with comparison to San Francisco Bay. In: Pederson J (ed) Marine Bioinvasions: Proceedings of the First National Conference, pp 130–138, January 24–27, 1999. Massachusetts Institute of Technology, Cambridge, Massachusetts

Molloy DP (1998) The potential for using biological control technologies in the management of *Dreissena* spp. Journal of Shellfish Research 17: 177–183

Murawski SA and Waring GT (1979) A population assessment of butterfish *Peprilus triacanthus*, in the northwestern Atlantic Ocean. Transactions of the American Fisheries Society 108: 427–439

Murphy NE and Goggin CL (2000) Genetic discrimination of sacculinid parasites (Cirripedia, Rhizocephala): implication for control of introduced green crabs (*Carcinus maenas*). Journal of Crustacean Biology 20: 153–157

Mutlu E (1999) Distribution and abundance of ctenophores and their zooplankton food in the Black Sea II. *Mnemiopsis leidyi*. Marine Biology 135: 603–613

Nagasaki K and Yamaguchi M (1998) Intra-species host specificity of HaV (*Heterosigma akashiwo* virus) clones. Aquatic Microbial Ecology 14: 109–112

Nagasaki K, Tarutani K and Yamaguchi M (1999) Growth characteristics of *Heterosigma akashiwo* virus and its possible use as a microbiological agent for red tide control. Applied and Environmental Microbiology 65: 898–902

Oviatt CA and Kremer PM (1977) Predation on the ctenophore, *Mnemiopsis leidyi*, by butterfish, *Peprilus triacanthus*, in Narragansett Bay, Rhode Island. Chesapeake Science 18: 236–240

Paine RT (1966) Food web complexity and species diversity. American Naturalist 100: 65–75

Perkins JH and Garcia R (1999) Social and economic factors affecting research and implementation of biological control. In: Bellows TS and Fisher TW (eds) Handbook of Biological Control, pp 993–1009. Academic Press, San Diego, California

Purcell JE and Cowan Jr JH (1995) Predation by the scyphomedusan *Chrysaora quinquecirrha* on *Mnemiopsis leidyi* ctenophores. Marine Ecology Progress Series 129: 63–70

Ross J and Johnson C (1998) Invasiveness and impact of the northern Pacific seastars *Asterias amurensis* on natural communities in SE Tasmania. In: Goggin CL (ed) Proceedings of a Meeting on the Biology and Management of the Introduced Seastar *Asterias amurensis* in Australian Waters, pp 13–17. Technical report no 15, Centre for Research on Introduced Marine Pests (CSIRO Division of Fisheries), Hobart, Australia

Schauff ME (1992) Systematics research in biological control. In: Kauffman WC and Nechols JR (eds) Selection Criteria and Ecological Consequences of Importing Natural Enemies, pp 15–25. Entomological Society of America, Lanham, Maryland

Secord D (2001) Symbioses and their consequences for community and applied ecology. In: Seckbach J (ed) Symbiosis: Mechanisms and Model Systems, pp 45–61. Kluwer Academic Publishers, Dordrecht, The Netherlands

Secord D and Augustine L (2000) Biogeography and microhabitat variation in temperate algal-invertebrate symbioses: zooxanthellae and zoochlorellae in two Pacific intertidal sea anemones, *Anthopleura elegantissima* and *A. xanthogrammica*. Invertebrate Biology 119: 139–146

Secord D and Kareiva P (1996) Perils and pitfalls in the host specificity paradigm. BioScience 46: 448–453

Settle WH, Ariawan H, Astuti ET, Cahyana W, Hakim A, Hindayana D, Lestart AS, Pajarningsih and Sartanto (1996) Managing tropical rice pests through conservation of generalist natural enemies and alternative prey. Ecology 77: 1975–1988

Shiganova TA and Bulgakova YV (2000) Effects of gelatinous plankton on Black Sea and Sea of Azov fish and their food resources. ICES Journal of Marine Science 57: 641–648

Simberloff D (1992) Conservation of pristine habitats and unintended effects of biological control. In: Kauffman WC and Nechols JR (eds) Selection Criteria and Ecological Consequences of Importing Natural Enemies, pp 103–117. Entomological Society of America, Lanham, Maryland

Simberloff D (2001) Eradication of island invasives: practical actions and results achieved. Trends in Ecology and Evolution 16: 273–274

Simberloff DS and Stiling P (1996a) How risky is biological control? Ecology 77: 1965–1974

Simberloff DS and Stiling P (1996b) Risks of species introduced for biological control. Biological Conservation 78: 185–192

Stanley JN and Julien MH (1999) The host range of *Eccritotarsus catarinensis* (Heteroptera: Miridae), a potential agent for the biological control of waterhyacinth (*Eichhornia crassipes*). Biological Control 14: 134–140

Strathmann RR (1990) Why life histories evolve differently in the sea. American Zoologist 30: 197–208

Stoecker DK (1992) 'Animals' that photosynthesize and 'plants' that eat. Oceanus 35: 24–27

Sweetman HL (1958) The Principles of Biological Control: Interrelation of Hosts and Pests and Utilization in Regulation of Animal and Plant Populations. William C. Brown Company, Dubuque, Iowa, 560 pp

Tarutani K, Nagasaki K and Yamaguchi M (2000) Viral impacts on total abundance and clonal composition of the harmful bloom-forming phytoplankton *Heterosigma akashiwo*. Applied and Environmental Microbiology: 4916–4920

Theisen BF and Jensen KR (1991) Genetic variation in six species of Sacoglossan Opisthobranchs. Journal of Molluscan Studies 57: 267–275

Thibaut T and Meinesz A (2000) Are the Mediterranean ascoglossan molluscs *Oxynoe olivacea* and *Lobiger serradifalci* suitable agents for a biological control against the invading tropical alga *Caulerpa taxifolia*? Comptes Rendus de l'Academie des Sciences (Paris) Life Sciences 323: 477–488

Thibaut T and Meinesz A (2001) Biological control of the invasive *Caulerpa taxifolia* in the Mediterranean Sea. Abstract Booklet, Second International Conference on Marine Bioinvasions, pp 138–139, New Orleans, 9–11 April 2001. Massachusetts Institute of Technology Sea Grant Program, Cambridge, Massachusetts

Thibaut T, Meinesz A, Amade P, Charrier S, DeAngelis K, Ierardi S, Mangialajo L, Melneck J and Vidal V (2001) *Elysia subornata* (Mollusca) a potential control agent of the alga *Caulerpa taxifolia* (Chlorophyta) in the Mediterranean Sea. Journal of the Marine Biological Association of the United Kingdom 81: 497–504

Thompson JN (1994) The Coevolutionary Process. University of Chicago Press, Chicago, 376 pp

Thresher RE (1996) Environmental tolerances of larvae of the European parasitic barnacle, *Sacculina carcini* (Rhizocephala: Sacculinidae): implications for use as a biological control agent against the European crab, *Carcinus maenas*. Technical report no 7, Centre for Research on Introduced Marine Pests (CSIRO Division of Fisheries), Hobart, Australia

Thresher RE (ed) (1997) Proceedings of the First International Workshop on the Demography, Impacts and Management of Introduced Populations of the European crab, *Carcinus maenas*. Technical report no 11, Centre for Research on Introduced Marine Pests (CSIRO Division of Fisheries), Hobart, Australia, 107 pp

Thresher RE (1999) Key threats from marine bioinvasions: A review of current and future issues. In: Pederson J (ed) Marine Bioinvasions: Proceedings of the First National Conference, pp 24–34, January 24–27, 1999. Massachusetts Institute of Technology, Cambridge, Massachusetts

Thresher RE, Werner M, Høeg JT, Svane I, Glenner H, Murphy NE and Wittwer C (2000) Developing the options for managing marine pests: specificity trials on the parasitic castrator, *Sacculina carcini*, against the European crab, *Carcinus maenas*, and related species. Journal of Experimental Marine Biology and Ecology 254: 37–51

Torchin ME, Lafferty KD, and Kuris AM (1996) Infestation of an introduced host, the European green crab, *Carcinus maenas*, by a symbiotic nemertean egg predator, *Carcinonemertes epialti*. Journal of Parasitology 82: 449–453

Trowbridge CD (1999) An assessment of the potential spread and options for control of the introduced green macroalga *Codium fragile* spp. *tomentosoides* on Australian shores. Consultancy report, CSIRO Marine Research/Centre for Research on Introduced Marine Pests, Hobart, Australia, 43 pp

Trowbridge CD and Todd CD (2001) Host-plant change in marine specialist herbivores: Ascoglossan sea slugs on introduced macroalgae. Ecological Monographs 71: 219–243

Turner E (1998) Groping in the dark! In: Goggin CL (ed) Proceedings of a Meeting on the Biology and Management of the Introduced Seastar *Asterias amurensis* in Australian Waters, pp 8–12. Technical report no 15, Centre for Research on Introduced Marine Pests (CSIRO Division of Fisheries), Hobart, Australia

Van Driesche RG and Bellows Jr TS (1996) Biological Control. Chapman & Hall, New York, 539 pp

Vinogradov ME, Shushkina EA, Musaeva EI and Sorokin PY (1989) Ctenophore *Mnemiopsis leidyi* (A. Agassiz) (Ctenophora: Lobata) – new settler in the Black Sea. Oceanology 29: 293–298

Waage JK (2001) Indirect ecological effects in biological control: the challenge and the opportunity. In: Wajnberg E, Scott JK and Quimby PC (eds) Evaluating Indirect Ecological Effects of Biological Control, pp 1–12. CABI Publishing, New York

Whayman D (1998) The industry perspective. In: Goggin CL (ed) Proceedings of a Meeting on the Biology and Management of the Introduced Seastar *Asterias amurensis* in Australian waters, pp 47–48. Technical report no 15, Centre for Research on Introduced Marine Pests (CSIRO Division of Fisheries), Hobart, Australia

Wickham DE and Roe P (1987) Selectivity in transmission to crab hosts by the symbiotic nemertean, *Carcinonemertes epialti*. Journal of Parasitology 73: 697–701

Williams S and Walker D (1999) Mesoherbivore–macroalgal interactions: feeding ecology of sacoglossan sea slugs (Mollusca, Opisthobranchia) and their effects on their food. Oceanography and Marine Biology: an Annual Review 37: 87–128

Wu M-Y, Hacker S, Ayres D, and Strong DR (1999) Potential of *Prokelesia* spp. as biological control agents of English cordgrass, *Spartina anglica*. Biological Control 16: 267–273

Zuljevic A, Thibaut T, Elloukal H and Meinesz A (2001) Sea slug disperses the invasive *Caulerpa taxifolia*. Journal of the Marine Biological Association of the United Kingdom 81: 343–344

Biological Invasions **5:** 133–141, 2003.

Did biological control cause extinction of the coconut moth, *Levuana iridescens*, in Fiji?

Armand M. Kuris
Department of Ecology, Evolution and Marine Biology, and the Marine Science Institute, University of California, Santa Barbara, CA 93106, USA (e-mail: kuris@lifesci.ucsb.edu; fax: +1-805-893-4724)

Received 10 September 2001; accepted in revised form 7 March 2002

Key words: Bessa remota, biological control, coconut moth, coconut palm, collateral damage, extinction, Fiji, host specificity, *Levuana iridescens*, non-target impact

Abstract

In 1925, J.D. Tothill and two colleagues set out to manage *Levuana iridescens*, the coconut moth of Fiji, using biological control. By 1930, they had succeeded so completely that this pest of the copra crop had been reduced to almost undetectable levels by the tachinid fly, *Bessa remota*, introduced from Malaya, and they had summarized their campaign in a thoroughly documented and well-illustrated monograph. The example of the coconut moth is presented in the modern literature as the first and best documented extinction of a species due to scientific biological control. The program has been severely criticized because the moth was unique, beautiful, and considered endemic to Fiji. Thus, this program is also portrayed as an example of the highly controversial practice of neoclassical biological control. However, a careful reexamination of this event discloses that the moth was likely not native to Fiji, appeared to be spreading through the Fijian Archipelago, and might have spread to other island groups in the South Pacific. Also, *L. iridescens* is probably not extinct. Collateral damage, that is, non-target impacts, did occur as native zygaenid moths have been attacked by the tachinid, and they may be extinct. The reasons for the control campaign of *L. iridescens* were not primarily economic. Tothill and colleagues were trying to protect copra so that ethnic Fijian culture, so dependent on the coconut palm threatened by *L. iridescens*, could be sustained. Hence, this control program represents a difficult clash of values: preservation of insect biodiversity *versus* preservation of indigenous Pacific Islander cultures. A strategy to search for *L. iridescens* populations is proposed and development of biological control of *B. remota*, using hyperparasitoids, is possible, but would require careful evaluation since it might release *L. iridescens* from suppression, have non-target impacts on native tachinids, and lack an economic motivation.

Introduction

As a generally cost-effective means of pest control, particularly compared to chemical control alternatives (pesticides, herbicides), biological control is clearly safe for humans. However, natural enemies, once they become established, are probably permanent faunal components. Hence, the impact of biological control agents on non-target native species is a cause for environmental concern (Louda et al. 1997; Follett and Duan 2000; Wajnberg et al. 2001; Henneman and Memmott 2001).

It is probably a safe generalization that native species are valued in most cultures. However, all are not valued equally (e.g., salmon or pandas vs. aphids and malaria), and pest status may be accorded some native species. Still, at least among environmentalists (and this probably includes most ecologists), collateral damage to non-target indigenous species poses a substantial and severe threat causing some critics to consider the use of biological control only as a last resort, or only to be used under very restricted circumstances, for the control of invasive species (Howarth 1991; Simberloff and Stiling 1996;

Louda et al. 1997; Cowie and Howarth 1998; Hager and McCoy 1998).

One historical example stands out as a worst case scenario. A biological control agent, scientifically evaluated, has been portrayed as the poster child for the great risk of the use of natural enemies for biological control of a pest. This example is the extinction of the coconut moth, *Levuana iridescens* Bethune-Baker, by a tachinid fly parasitoid, *Bessa* (= *Ptychomyia*) *remota* (Aldrich). This example was recognized by Roberts (1986) and highlighted by Howarth (1991), and his perspective has been repeatedly cited in the recent cautionary literature (e.g., Simberloff and Stiling 1996; Barratt et al. 2000; Lynch et al. 2001; Holt and Hochberg 2001). Howarth made the following case:

> The control of the coconut moth, *L. iridescens*, on Fiji by the purposeful introduction of the tachinid fly *Bessa remota* from Malaysia in 1925 was described in great detail (Tothill et al. 1930) and is often cited as a classic example of biological control (DeBach 1974; Rao et al. 1971; Sweetman 1958). The last authentic specimen of this endemic monotypic genus of Zygaenidae was collected in 1929 (Robinson 1975; Tothill et al. 1930), although the species may have survived until the 1940s (Roberts 1986; Robinson 1975). The species, a widespread local pest, became endangered in less than 2 years, *and its demise is probably the best documented study of extinction among the insects* [italics mine] (Tothill et al. 1930). Another unrelated zygaenid, *Heteropan dolens*, was extirpated from Fiji at the same time (Robinson 1975). The impacts on other non-pestiferous native Fijian Lepidoptera were not recorded, even though the fly still occurs on Fiji, parasitizing non-target species (Russell 1986). Tothill et al. (1930) were unable to find the major alternative hosts in Fiji.

To this, I will add that *L. iridescens* was a distinctive species with some unique biological attributes (Tothill et al. 1930). From my personal aesthetic perspective, it was also a beautiful species, colored an iridescent dark blue (Figure 1).

Fortunately, the campaign against the coconut moth in Fiji has been described in eloquent detail by the Canadian entomologist J.D. Tothill and his two young British graduate assistants, T.H.C. Taylor and R.D. Paine who undertook the assignment and reported their studies in a monographic report on the coconut moth, and the history of its control by means of the

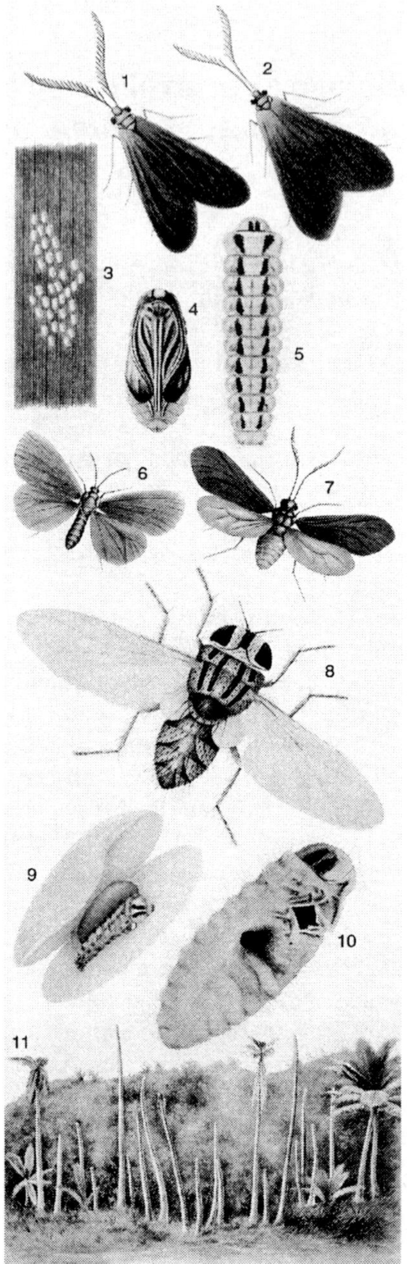

Figure 1. (1) *L. iridescens*, male, resting, (2) *L. iridescens*, female, resting, (3) *L. iridescens* egg cluster on coconut palm leaf, (4) *L. iridescens*, pupa, (5) *L. iridescens*, fifth instar larva, (6) *H. dolens*, wings spread, (7) *L. iridescens*, female, wings spread, (8) *B. remota*, adult, (9) cocoon of parasitized *L. iridescens*, opened to show dead larva of *L. iridescens* and pupa of *B. remota*, (10) larval corpse of *L. iridescens* with exit hole for *B. remota*, (11) death of coconut palms on Ovalau, after successive outbreaks of *L. iridescens*, prior to control by *B. remota*. (From Tothill et al. [1930]; paintings of insects by H.W. Simmonds; painting of shoreline by W.J. Belcher.)

parasitoid, *B. remota*, to the Government of Fiji and the Imperial Bureau of Entomology. Some invaluable additional comments are provided in a memoir by Paine (1994), who stayed on in Fiji until 1936, and then returned, as an entomologist, from 1956 to 1966.

Much has changed from the time of the pest control efforts of Tothill and his colleagues (1925–1930). In addition to the many technological advances, values on matters such as damage to native insects, have changed; often in ways difficult to discern. It is because of these social aspects that we are particularly indebted to Tothill, Taylor and Paine, for they very carefully specified their motivation for undertaking their control efforts. They were also conscious of some of the issues that provide us a perspective bridging the past 75 years.

To evaluate what happened to the coconut moth, and consider whether a comparable campaign would be appropriate in the environmental climate of today, five matters must be addressed:

1. Was the coconut moth actually endemic to Fiji?
2. Did collateral damage occur?
3. Is the coconut moth actually extinct?
4. Why did they attempt to control the coconut moth?
5. Was the control campaign conducted with human and environmental safety in mind?

The campaign against the coconut moth in Fiji

Was the coconut moth an endemic species?

Tothill et al. (1930) recognized that evaluating the zoogeography of *L. iridescens* would be critical to the development of a successful biological control strategy. The most effective biological control agents generally come from the region of origin of the pest, and prospecting for useful natural enemies is conducted in this region. The practice of detection of natural enemies in the region of origin of the pest, followed by their importation, evaluation and release, is termed 'classical' biological control (Smith 1948; Garcia et al. 1988).

The determination of the native zoogeographic status of *L. iridecens* is also key to a modern evaluation of the conduct of the coconut moth biological control campaign. If the moth were native to Fiji, then importation of a natural enemy form elsewhere would be an example of 'neoclassical' biological control (Lockwood 1993). Hokkanen and Pimentel (1984, 1989) have argued for the use of such 'new associations' precisely because they are not co-evolved, and the natural enemy may have a great impact on a 'naïve' pest. However, as Lockwood (1993) stresses, neoclassical biological control agents are selected *a priori* because they have low host specificity. Hence, their potential impacts on non-target species are potentially severe, and unbounded, with unintended consequences difficult to predict.

If the coconut moth were not native to Fiji, this would explain why it had recently become abundant and caused such severe economic and ecological problems. Weighing the evidence, Tothill et al. (1930) decided that *L. iridescens* was probably *not* a Fijian native endemic; but, more likely, represented a cryptic invasion from the Malayan/Melanesian regions to the west of Fiji (Figure 2). They based this conclusion on five points.

First, the zoogeography of *L. iridescens* in Fiji was odd. Unlike most Fijian endemics, it was not found widely among the main islands of the archipelago, but only on Viti Levu and a few small islands near its coast (Figure 2).

Second, extensive dissections of *L. iridescens* found it to be unparasitized. This was very unusual, as other zygaenid lepidopterans, especially its closest relatives, were frequently attacked by parasitoids. In contrast, introduced species are commonly unparasitized, and classical biological control aims to re-associate missing upper trophic level organisms with pestiferous exotic target species.

Third, the reason it was a serious pest of coconut trees was that it had repeatedly exhibited population outbreaks – periods of great abundance, causing defoliation and even death of many trees over long stretches of coast (Figure 1). But, these outbreaks had never been observed prior to 1877, during the time of the sandlewood trade that produced considerable shipping traffic between Fiji and the east coast of Asia. After that time, outbreaks of *L. iridescens* became regular occurrences. Coastal coconut trees on Viti Levu experienced repeated trunk constrictions associated with the greatly slowed growth that results from the repeated defoliation episodes caused by *L. iridescens*. Although trunk constrictions can be caused by other growth inhibiting factors, the greatly increased frequency of trunk constrictions on trees after the 1870s confirms the lack of reports of outbreaks in the early colonial and pre-colonial periods.

Fourth, *L. iridescens* relatives in Malaya, Java, New Guinea and the Solomon Islands are generally rare species with infrequent localized sporadic outbreaks.

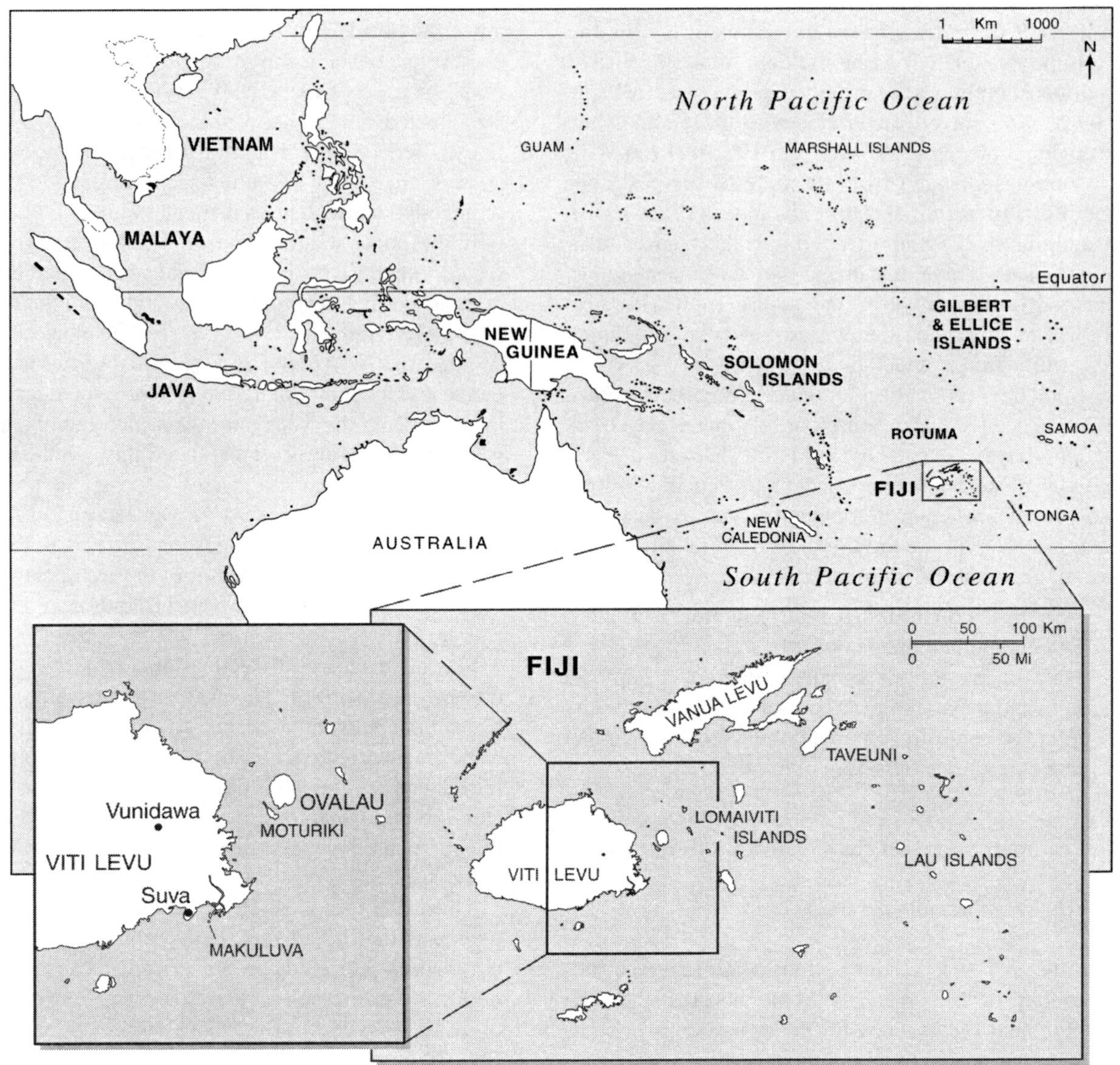

Figure 2. Places named in text are shown with historic names on a map of the Western Pacific with nested insets of the Fiji Archipelago, and eastern Viti Levu with offshore islands. Spelling of Fijian locations is modernized.

Hence, the outbreaks of *L. iridescens* from 1877 on are unusual life history features of these zygaenids, whose abundances are presumably generally held in check by natural enemies in their native ranges.

Fifth, *L. iridescens* had recently spread from Viti Levu to Ovalau and Moturiki (Figure 2), devastating copra plantations on those islands. Such an explosive range expansion is characteristic of recently established invaders. It also strongly suggested that other islands were at risk of coconut moth invasion and establishment.

While these points, particularly in the aggregate, support the hypothesis that *L. iridescens* was not, in fact, native to Fiji, they do not disprove the alternative that it was endemic. A few native endemic species do have odd restricted distributions within the archipelago. A search for *L. iridescens* in Malaya, Java and New Guinea did not detect it there. But, that is a lot of territory to search for a possibly rare species, and its recent outbreaks in Fiji might be caused by unknown changes in cultivation practices for copra (Tscharntke 2000).

Evaluating the possibilities, Tothill et al. (1930) concluded that *L. iridescens* was most likely not native to Fiji. Hence, they assumed they were working in the mode of classical biological control, rather than the more contentious (Hokkanen and Pimentel 1989; Lockwood 1993) neoclassical biological control strategy (using introduced natural enemies against native pests).

The issue of the endemicity of the coconut moth is central to an evaluation of its putative eradication as a tragedy. The conclusion remains elusive. Kalshoven (1981) considered *L. iridescens* to be conspecific with *Cathartona* (=*Artona*) *catoxantha* (Hampson), a widespread Indo-Malayan species (and the source host for *B. remota*). However, this taxonomic assignment is considered unlikely (Sands 1997).

Did collateral damage occur?

Since *L. iridescens* was the target of a classical biological control program, reduction of its population densities to undetectable levels was deliberately planned. Only its extinction might be regarded as 'collateral'. However, the parasitoid, *B. remota*, is a low host specificity natural enemy (unlikely to be released in the current regulatory and cultural climate of some Western countries (Sands 1997)). Tothill et al. (1930) reported that it readily attacked other zygaenid moths native to Fiji; one of which, *H. dolens* Druce, is now rare and may be extinct (Robinson 1975; Howarth 1991). So, collateral damage to non-target native species from *B. remota* clearly occurred. It is important to note that, to date, no comprehensive survey of native zygaenid moths for *B. remota* parasitism has been undertaken, as has recently been conducted for Hawaiian biological control agents attacking native species (Henneman and Memmott 2001). It would be valuable to also quantify the nature and extent of the collateral damage to native Fijian Lepidoptera in this influential case.

Over the past 75 years, other changes have come to Fiji that could have had a additive negative impact on abundance of *L. iridescens*, and the other related native moths that serve as hosts for the *B. remota* (Sands 1997). For example, coconut plantations have displaced native palms over most of the lowland regions of Viti Levu. This may have reduced habitat for these moths and could have decreased access to possible host plant refugia for the coconut moth. Pesticide use to control pest insects may also have negatively affected these zygaenids.

Is L. iridescens *actually extinct?*

Despite Tothill et al.'s (1930s) suspicions, and Howarth's (1991) claim, that the last specimen was collected in 1929, the coconut moth may still persist in Fiji. A specimen was collected in 1941 (Sands 1997). More importantly, Paine (1994) reported an outbreak of *L. iridescens* (soon followed by *B. remota* parasitism) on coconut palms near Vunidawa in 1956 (Figure 2). Since Paine was one of the original participants in the *L. iridescens* campaign, responsible for rearing the moths and the parasitoids, his identification of *L. iridescens* in 1956 must be considered highly reliable. Hence, its pattern of long periods of rarity, with occasional sporadic outbreaks, now resembles the temporal pattern of abundance of its relatives, such as *C. catoxantha*, on other palms in Java (Sands 1997).

Many exotic species go through an initial population outbreak phase, but then decline, sometimes below pestiferous levels, sometimes even to extinction. In some cases, this has been attributed to apparently successful biological control (Beirne 1975). However, the 50 year record of outbreaks of *L. iridescens* in Fiji, without a hint of decline (that is why the call for its control came in the 1920s), followed by its collapse to perhaps eradication, with very high levels of parasitization by *B. remota*, indicate that we can reject the hypothesis of decline to extinction due to causes other than its control by *B. remota*.

Why was L. iridescens *the target of a biological control campaign in Fiji?*

To assess the unquantified significance of the non-target impacts, and to evaluate the risks associated with this venture, it is necessary to appreciate the motivation for the control program against *L. iridescens*. Fortunately, the reasons are very clearly specified by Tothill et al. (1930) and are somewhat surprising for such an early agricultural pest control campaign. Tothill et al. (1930) note that the principle economic crop of Fiji was sugar cane. It used immigrant Indian labor and was an export crop important to colonial planters, merchants and the Government. However, ethnic Fijians would be little affected by changes in the cane economy. However, they crucially relied on copra for subsistence, many cultural practices and as a source of cash. The loss of copra was seen, by the colonial Administration, as tantamount to the demise of Fijian culture. Tothill et al. (1930) remark on these matters extensively and with clarity. Other than their archaic use of some terms, their views are quite modern. They quote Sir Morris Hedstrom, Senior Member of the Legislative Council:

> Any serious check to our sugar industry would have a disastrous effect on the Colony, particularly upon the Administration and the mercantile community, but, speaking broadly, it would not have a crushing effect on the native race, whereas the loss of the copra industry would utterly impoverish the natives of Vanua Levu, Taviuni, Lomaiviti, and Lau, and render conditions of life and Administration of those districts very difficult.
>
> Owing to the central position of Ovalau and Moturiki, and the constant traffic from those islands to Vanua Levu, Taviuni and Lau, it would be unsafe to reckon on any definite period of immunity for the other coconut-bearing areas in the group.

Tothill et al. (1930) add:

> Another respect in which coconuts are more important to this Colony than cane is that from time immemorial the coconut palm has been a mainstay of the Fijian race. In times of failure of taro and yams the coconut has provided meat and in times of drought has provided drink for populous villages in outlying islands. In a degree the tree has been the dairy cow of the Native Race and has enabled healthy children to be raised without the aid of the domestic cow. Thus the coconut is the chief source of material wealth – of ready cash – of the Fijians, and it enables islands to be inhabited and prosperous that would otherwise be unpopulated and barren.
>
> Finally, the wealth derived from coconut palms remains largely in the Colony, while part of that derived from sugar cane goes out of the Colony to shareholders of a Company that has headquarters in Sydney.

Hence, they conclude:

> We are inclined to think that the most important result of the campaign is the one least capable of measurement, namely, that coconuts can now be grown on Viti Levu.

Tothill et al. (1930) also recognized that more was at stake than just the preservation of ethnic Fijian culture on Viti Levu. They expressed a prescient concern for the future spread of *L. iridescens* and its consequent potential to impact other Pacific island cultures, since it had already spread to the neighboring islands of Viti Levu.

> The alarm was caused by *L. iridescens* breaking its traditional bounds on the large island of Viti Levu and finding its way to Ovalau and neighbouring islands. It had been present on Viti Levu for so many years that coconuts were there not regarded as a source of wealth.
>
> Another result of the campaign, also not calculable in money terms, is that it helps to ensure the prosperity of the native races of the South Pacific depending so largely upon the coconut palm. *Levuana* introduced without parasites to the Lau islands of Fiji, into Rotuma or into the Gilbert and Ellice group and into many other islands of Polynesia and Micronesia might have caused these islands gradually to become depopulated, and, as transportation by sea became more rapid and more frequent, such introductions might have occurred.

The islands of the Pacific, particularly the atolls, are deficient in resources, have depauperate floras, and can sustain only very limited agriculture. Copra, with all its uses, is the crop that permits indigenous subsistence cultures to flourish on those islands. Expansion of the range of *L. iridescens* due to increased boat traffic – a concern expressed by these scientists in 1930! – seems a foregone conclusion. It appears that the near eradication of *L. iridescens* by Tothill, Taylor and Paine prevented a probable cultural catastrophe from occurring had the copra industry collapsed because of *L. iridescens*.

The biology of the outbreaks on Viti Levu and then on Ovalau and Moturiki support these worst case concerns. Tothill et al. (1930) document massive and repeated defoliation of coconut trees, reducing their growth considerably and often killing even large trees. Their illustrations show long coastal views with nearly all coconut palms showing all but the newest leaves dead.

How were human health and environmental safety concerns evaluated during the Levuana *campaign?*

The control program was carefully developed with a concern for environmental safety. As they attempted a variety of control procedures Tothill et al. (1930)

clearly demonstrated an Integrated Pest Management (IPM) philosophy 40 years before the IPM concept was explicitly developed. In the 1920s, few chemical pesticides were available. Only arsenate of lead exhibited efficacy, and spraying this up into coconut palms was tested (using a fire hose pump and nozzle). This was rejected as unsafe (and costly). The use of generalist bird predators was proposed by avian enthusiasts, and this was vigorously (and successfully) opposed by Tothill et al. (1930). They were also careful to not introduce other pests (recall the parasitoid was isolated from *C. catoxantha*), and they also carefully cultured their initial stocks to ensure that no hyperparasitoids were introduced.

Conclusions

It is uncertain whether *L. iridescens* was native to Fiji. It also may not be extinct. So, the repeatedly cited story of the extinction of *L. iridescens* by a biological control agent as a worst-case scenario is also uncertain, perhaps even unlikely. Collateral damage to other moths did certainly occur, but the extent and severity of the non-target impacts have not been quantified. How great a loss this is depends on whether these native zygaenids are, in fact, extinct.

It seems probable that there would be further host-specificity testing of potential control agents today, and other, perhaps more specific control agents than *B. remota* would be selected.

The primary goal of the biological control campaign against *L. iridescens* was to protect the cultures of indigenous Fijian, and other South Pacific island peoples. In this it was spectacularly successful. Historically, the campaign against the coconut moth can be viewed as an antecedent to the gardenification concept. According to Janzen (1998), humans have had an overwhelming impact on the biota of the world, and it is our responsibility to actively manage it in perpetuity to the best of our ecological ability. To do this requires explicit consideration of values, costs and benefits. In this spirit, Tothill et al. (1930) attempted to preserve biodiversity and minimize impacts on human health. So, to evaluate this history from the vantage point of today forces a poignant clash of values: insect biodiversity *versus* indigenous culture. In my opinion, reasonable people may differ as to which value has the greater merit.

Future work

While this review recounts history, the story is not yet complete. There is something to consider and something to do.

If the *H. dolens*, other native zygaenids, and perhaps *L. iridescens* are not actually extinct, but have become so rare as to be threatened with extinction by *B. remota*, and this loss of biodiversity is judged to be excessive collateral damage to non-target native species, a remedy is potentially available. It is possible that *B. remota* itself might be brought under biological control. Tothill et al. (1930) remark that hyperparasitoids (= secondary parasites) were common in *B. remota* collected in Malaya, and that these hyperparasitoids 'detract considerably from the efficiency of *B. remota* in Malaya, and had these been allowed to escape in Fiji the victory over *Levuana* might have been much less prompt and decisive than it had proved to be'. Of course, such a consideration would itself require careful evaluation and safety tests. A reduction in the efficacy of *B. remota* as a natural enemy might boost *L. iridescens* populations and those of the native zygaenids that are not already extinct. Of course, it might also risk the effective economic control of a serious copra pest. Collateral damage of a hyperparasitoid release to native Fijian tachinids would also have to be evaluated.

Use of a biological control agent to remedy impact of a biological control agent will likely strike advocates on both sides of this contentious issue as unwarranted. However, as a policy modeling exercise, it would provide an evaluation of the relative weights of the values that are in conflict in the history of the biological control of the coconut moth in Fiji.

Finally, a rationally designed search should be conducted in Fiji to seek the coconut moth in potential refugia. Tothill et al. (1930) considered that the remarkable efficacy of *B. remota* as a biological control agent for *L. iridescens* was primarily due to the strong flying ability of the tachinid compared to the very poor flight dispersal ability of the coconut moth. Likely places to seek relict populations of *L. iridescens* would be small offshore islands with limited floras and few alternative lepidopteran hosts for *B. remota*. Tothill et al. (1930) specifically noted that, as *L. iridescens* became vanishingly rare, the last numerous populations could be found on small leeward islands, such as Makuluva in the delta of the Rewa River. Prior to biological control by *B. remota*, *L. iridescens* was rare on windward islands and abundant on leeward islands. A search for

L. iridescens elsewhere in the western Pacific might also be fruitful.

Sands (1997) noted that *Levuana* also fed on other species of native palms. These species should also be examined for *L. iridescens* and also *H. dolens*.

To determine if *L. iridescens* and the less well known endemic *H. dolens*, are extinct in Fiji, a coherent search for them should be initiated. Its design should be based on elements of the ecology of *L. iridescens* and *B. remota* that may point to refugia of the former from the latter species. For example, Tothill et al. (1930) note the common occurrence of *L. iridescens* on several other species of palms. The key ecological attribute of *B. remota* that promoted its spectacular efficacy against *L. iridescens* is its much greater dispersal capacity. *B. remota* is a strong flier, while *L. iridescens* is a poor flier (Tothill et al. 1930), supporting that moth's near absence on the windward side of Viti Levu and of small islands on the leeward coast.

A dedicated searching strategy for *L. iridescens* should include:

(1) The discovery of an outbreak of *L. iridescens* at a copra plantation near Vunidawa suggests that this may be near a possible refugium for *L. iridescens* and perhaps for *H. dolens*. Vunidawa is deep in the interior of Viti Levu. This sort of location may offer a more diverse palm flora for zygaenid moths than does the maritime flora associated with naturally distributed coconut palms. This, and other similar locations, should be explored with considerable attention paid to the several other native palm species known to be hosts for *L. iridescens* and related zygaenids (Tothill et al. 1930).

(2) Small leeward islands should be explored for *L. iridescens*. Before the release of *B. remota*, *L. iridescens* was most abundant on the small islands too the east of Viiti Levu, and on those islands its populations were concentrated on the leeward side, suggesting that these moths had limited ability to disperse windward. After the release of *B. remota*, these leeward islands retained some of the last abundant populations of *L. iridescens* suggesting that *B. remota* had some difficulties reaching them, or sustaining its populations on them (perhaps because alternate hosts were less available on these small depauperate islands. Some of these small islands are uninhabited or little visited (Kuris, pers. obs.).

(3) In order to find a rare and patchily distributed organism, leeward and interior areas should be explored extensively, looking at relatively few host plants from many sites, rather than intensively searching many plants at relatively few (the most probable) sites.

Acknowledgements

I thank Kevin Lafferty, Mark Torchin and James Carlton for discussions of the *Levuana* story, and Bill Reece, and other Fijians on the island of Galoa, for describing to me the importance of the coconut palm to South Pacific islanders. Mark Rigby, Mark Hoddle, and an anonymous review made valuable comments on the manuscript. This research was funded by a grant from the National Sea Grant College Program, National Oceanographic and Atmospheric Administration (NOAA), US Department of Commerce under grant number NA06RG0142, project number R/CZ-162 through the California Sea Grant college system, and in part, by the California State Resources Agency. The views expressed herein are those of the author and do not necessarily represent the views of NOAA or any of its sub-agencies. The US government is authorized to reproduce and distribute this paper for governmental purposes.

References

Barratt BS, Goldson SL, Ferguson CM, Phillips CB and Hannah DJ (2000) Predicting the risk from biological control agent introductions: a New Zealand approach. In: Follett PA and Duan JJ (eds) Nontarget Effects of Biological Control Agents, pp 59–75. Kluwer Academic Publishers, Boston

Beirne BP (1975) Biological control attempts by introductions against pest insects in the field in Canada. Canadian Entomologist 107: 225–236

Cowie RH and FG Howarth (1998) Biological control: disputing the indisputable. Trends in Ecology and Evolution 13: 110

DeBach P (1974) Biological Control by Natural Enemies. Cambridge University Press, London, 323 pp

Follett PA and Duan JJ (2000) Nontarget Effects of Biological Control Agents. Kluwer Academic Publishers, Boston, 316 pp

Garcia R, Caltagirone LE and Gutierrez AP (1988) Comments on a redefinition of biological control. BioScience 38: 692–694

Hager HA and McCoy KD (1998) The implications of accepting untested hypotheses: a review of the effects of purple loosestrife (*Lythrum salicaria*) in north America. Biodiversity and Conservation 7: 1069–1079

Henneman ML and Memmott J (2001) Infiltration of a Hawaiian community by introduced biological control agents. Science 293: 1314–1317

Holt RD and Hochberg ME (2001) Indirect interactions, community modules and biological control: a theoretical perspective.

In: Wajnberg E, Scott JK and Quimby PC (eds) Evaluating Indirect Ecological Effects of Biological Control, pp 13–37. CABI Publishing, Wallingford, UK

Hokkanen H and Pimentel D (1984) New approach for selecting biological control agents. Canadian Entomologist 116: 1109–1121

Hokkanen H and Pimentel D (1989) New associations in biological control: theory and practice. Canadian Entomologist 121: 829–840

Howarth FG (1991) Environmental impacts of classical biological control. Annual Review of Entomology 36: 485–509

Janzen D (1998) Gardenification of wildland nature and the human footprint. Science 279: 1312–1313

Kalshoven LGE (1981) Pests of Crops in Indonesia, revised and translated by Van der Laan PA and Rothschild GHL. PT Ichtiar Baru-Van Hoese, Jakarta, Indonesia, 701 pp

Lockwood JA (1993) Environmental issues involved in biological control of rangeland grasshoppers (Orthoptera: Acrididae) with exotic agents. Environmental Entomology 22: 503–518

Louda SM, Kendall D, Connor J and Simberloff D (1997) Ecological effects of an insect introduced for the biological control of weeds. Science 277: 1088–1090

Lynch LD, Hokkanen HMT, Babendreier D, Bigler F, Burgio G, Gao Z-H, Kuske H, Loomans A, Menzler-Hokkanen I, Thomas MB, Tommasini G, Waage, JK, van Lenteren JC and Zeng Q-Q (2001) Insect biological control and non-target effects: a European perspective. In: Wajnberg E, Scott JK and Quimby PC (eds) Evaluating Indirect Ecological Effects of Biological Control, pp 99–124. CABI Publishing, Wallingford, UK

Paine RW (1994) Recollections of a Pacific Entomologist 1925–1966. Australian Centre for International Agricultural Research, Canberra, 126 pp

Rao VP, Ghani MA, Sankaram T and Mathur KC (1971) A Review of the Biological Control of Insects and Other Pests in South-East Asia and the Pacific Regions. Commonwealth Agricultural Bureau, London, 149 pp

Roberts LIN (1986) The practice of biological control – implications for conservation, science and the community. Weta Bulletin of the Entomological Society of New Zealand 9: 76–84

Robinson GS (1975) Macrolepidoptera of Fiji and Rotuma. Classey, Faringdon, UK, 362 pp

Russell DA (1986) The role of the entomological society in insect conservation in New Zealand. Weta Bulletin of the Entomological Society of New Zealand 9: 44–54

Sands DPA (1997) The safety of biological control agents: assessing their impact on beneficial and other non-target hosts. Memoirs of the Museum of Victoria 56: 611–616

Simberloff D and Stiling P (1996) Risks of species introduced for biological control. Biological Conservation 78: 185–192

Smith HS (1948) Biological control of insect pests. In: Reuther W, Calaran EC and Carman GE (eds) Biological Control of Citrus Insects. The Citrus Industry Vol IV, pp 276–320. University of California, Division of Agricultural Sciences Publications, Berkeley, California

Sweetman HL (1958) The Principles of Biological Control. Brown, Dubuque, Iowa, 560 pp

Tothill JD, Taylor THC and Paine RW (1930) The Coconut Moth in Fiji: a History of Its Control by Means of Parasites. Imperial Bureau of Entomology, London, 269 pp

Tscharntke T (2000) Parasitoid populations in the agricultural landscape. In: Hochberg ME and Ives AR (eds) Parasitoid Population Biology, pp 235–2253. Princeton University Press, Princeton, New Jersey

Wajnberg E, Scott JK and Quimby PC (eds) (2001) Evaluating Indirect Ecological Effects of Biological Control. CABI Publishing, Wallingford, UK, 261 pp

Author index